THE SKY ISN'T EMPTY:

America's Blind Spot in the Age of Drones

"Awareness without action is chaos. Action without awareness is oblivion."

CSM Sheldon A. Watson

We are the watchers, not bystanders. We are the stewards of liberty, not its consumers. We do not inherit freedom; we maintain it. Not through comfort, but through vigilance. Not through silence, but through accountability.

We are the sentinels of a connected nation. Our tools are innovation. Our duty is restraint. Our strength lies not in dominance, but in discernment.

We do not outsource conscience. We embody it.

In every age, the sky has held more than clouds; it holds our choices, our consequences, our calling.

And in this age of drones and data. We must choose to *see*, not just to watch.
To *act*, not just to automate.

It's waiting.

S.A.W.

The Sky Isn't Empty: America's Blind Spot in the Age of Drones

By

© 2025 S.A.W. Legacy Group LLC

Published by SAW Legacy Press, an imprint of S.A.W. Legacy Group LLC,

Colorado Springs, Colorado

ISBN: _____

Printed in the United States of America

Domain: www.SAWLegacyGroup.com

Email: info@SAWLegacyGroup.com

TABLE OF CONTENTS

Preface

The Sky Isn't Empty: America's Blind Spot in the Age of Drones

"The most dangerous enemy is the one we do not see." That line, often repeated and rarely attributed, carries more weight today than ever before. The first time I witnessed a drone drop an explosive payload, it wasn't in a movie or a classified briefing. It came through a grainy feed from a war zone thousands of miles away. The target wasn't a tank or a hardened position. It was a soldier, standing beside a vehicle, checking his gear. The drone, a cheap, off-the-shelf quadcopter available online for less than $500, hovered silently overhead, invisible to the naked eye. Then, with a mechanical twitch, it released its payload. The screen flashed white. One soldier vanished. One more headline emerged in a war that had changed forever.

Years have passed since that moment, but what lingered wasn't the explosion; it was the silence. No radar detected it. No alarm sounded. No defense system responded. The sky, once a symbol of superiority and sanctuary, had shifted beneath our feet. Today, that same type of drone can launch from an apartment balcony, glide over a stadium, buzz a substation, or loiter above a schoolyard in any American city. And most of the time, no one even looks up.

After nearly three decades in uniform, I've learned that threats evolve faster than policies, and fear spreads faster than facts. Yet despite all our planning, training, and defending, we continue to overlook the most obvious truth: the sky above us no longer belongs exclusively to us. When I began studying unmanned systems and the policies that attempt to govern them, I didn't just uncover technical gaps; I also uncovered a psychological one. Our greatest vulnerability isn't buried in circuitry or cloaked in code; it's embedded in our assumptions. We still believe danger comes from "over there," not overhead. That belief isn't just outdated, it's dangerous.

This book is not intended to stoke fear. It exists to provoke awareness. The drone problem isn't confined to military installations or federal agencies. It's an American problem, one that spans backyards, ballparks, boardrooms, and beyond. The age of drones has arrived, and it's not waiting for legislation to catch up. For every sophisticated military-grade UAV, thousands of commercial and hobbyist drones now share the same airspace. Some are innocent. Some are curious. A few are malicious. The challenge isn't just the drone itself; it's the technology's dual-use nature. What delivers medication today can deliver a bomb tomorrow. What films your wedding, can also scout your power grid. And what entertains your child can, in the wrong hands, surveil your neighborhood.

However, the bigger challenge, and the greater opportunity, lies not in the machines, but in the mindset of the people beneath them. That means **YOU**. Our national defense no longer begins with a uniform or a badge; it starts with awareness. Homeland Defense has evolved. Vigilance, ethics, and civic responsibility now define it. "See something, say something" isn't just a slogan; it's the first line of aerial defense in a nation where every citizen lives beneath the same vulnerable sky.

Accordingly, this book serves as a civilian translation of a complex truth: drones aren't coming; they're already here. And while policymakers debate, the airspace fills. Over the following fifteen chapters, you'll encounter the realities of the drone revolution: the innovation, the risks, the policy paralysis, and the moral crossroads we now face. You'll learn how a $100 device can cripple a $10 million system. You'll see how artificial intelligence and autonomy are rewriting the rules of warfare, while bureaucracy struggles to keep pace. But you'll also discover something far more potent than any drone: grassroots action. Local awareness, responsible innovation, and citizen engagement remain our most effective countermeasures to these challenges.

This isn't just a book about technology. It's a book about responsibility. Because drones aren't going away. However, neither is the American spirit one of adapting, protecting, and overcoming. The sky isn't empty. And neither are we.

Introduction

The Age of Unseen Enemies

From above, America looks peaceful, with cities glowing like constellations, rivers of headlights flowing through interstates, and small towns tucked under blankets of quiet. But from 400 feet and below, a very different picture emerges.

At that altitude, beyond the reach of most radars and beneath the authority of traditional air defense, an entire ecosystem of unmanned aircraft moves quietly, constantly, and largely unmonitored. These are the new occupants of the American sky: drones delivering packages, filming real estate, mapping crops, inspecting towers, and in some cases conducting reconnaissance for those with darker intentions.

It's a world most people don't see, and fewer still understand.

From Hobby to Hazard

In 2006, the FAA recorded just a few hundred registered drones in U.S. airspace. By 2025, that number surpassed 1.8 million. The same flight technology that once required a hangar and a pilot's license can now fit in a backpack and be flown with a smartphone.

Technology democratized flight, but it also brought unintended consequences. The barrier to entry is gone; so too is the barrier to misuse.

We've seen drones interfere with wildfires, delay commercial flights, spy on neighbors, and disrupt prisons. Overseas, they've dropped explosives on tanks and ammunition depots. Domestically, the same tools used to protect lives can be repurposed to take them.

The Problem We Don't Want to Admit

America has built systems to detect missiles that travel thousands of miles; however, it has not built systems to detect drones that travel a few hundred feet. We've also created sophisticated cybersecurity frameworks, but struggle to regulate the airspace above our own neighborhoods.

Part of the problem is cultural. Americans still see drones as gadgets, not weapons. We laugh at YouTube videos of drone failures but miss the message: these machines are rapidly evolving, and we're standing still.

Our response gap isn't just technological; it encompasses legislative, moral, and communal aspects. And in that gap lies our danger.

The Opportunity Hidden in the Threat

Yet for all the challenges, there's a silver lining. Because the very thing that makes drones dangerous, their accessibility, also makes them a tool for empowerment.

With proper education and civic awareness, the same technology that can harm can also be used for protection. Communities can utilize drones for search and rescue operations, environmental monitoring, and local security purposes. Farmers, engineers, and first responders are already proving that responsible drone use can strengthen national resilience.

But that requires one shift: awareness must replace apathy.

The American people cannot be spectators. Security isn't a service we outsource; it's a shared responsibility.

A Shared Horizon

In this book, I'll take you through the layers of this modern challenge, from the battlefield lessons of Ukraine to the regulatory struggles in Washington, from the vulnerabilities of our power grid to the moral questions of surveillance and privacy.

Each chapter builds toward one conclusion: the defense of the homeland begins not above us, but among us.

While drones fly high, the responsibility for our security rests firmly on the ground, with *us*.

Dedication

To Soleil, Silas, and Shane, My most significant sources of light, my anchor in every storm, and my reason for looking up when the world feels heavy. Your love and patience fill the sky I once thought was empty.

Acknowledgments

This book would not exist without the people who shaped my path, sharpened my purpose, and steadied my hands when the mission grew long.

To my wife, Soleil, thank you for being the constant rhythm in my life, the calm behind every decision, and the faith that reminded me who I am when titles fade.

To my sons, Silas and Shane, your curiosity, humor, and courage have kept me grounded while I tried to understand the world above us.

To my mentors, peers, and the men and women of the Armed Services, thank you for showing me that leadership begins not with command, but with character.

To my professors and colleagues at Liberty University, your insistence on pairing faith with scholarship helped me find my voice between doctrine and duty.

And to those who have worn the uniform, past, present, and future, this book is for you. For the moments you stood watch when no one noticed, for the missions no one will ever know, and for the quiet prayers that carried you through them.

Author's Note

This book began not with an idea, but with a question that refused to leave me: "What happens when the sky becomes both battlefield and mirror?"

I've spent over two and a half decades in uniform, as a tanker, leader, advisor, and now Command Sergeant Major. I've seen war in the desert, peace on uneasy terms, and innovation that outpaced our ethics. But I've also seen what faith, family, and purpose can restore when technology tempts us to forget the human heartbeat behind every decision.

The Sky Isn't Empty is more than analysis. It's a reflection, part field manual, part confession, written by a soldier who has watched the horizon change and refuses to look away.

This is not a book about fear. It's about stewardship, freedom, responsibility, and of the moral compass that guides us when data replaces direction.

If the sky reflects the soul of a nation, then our task is simple: to make sure what it shows is still worth defending.

SAW

Foreword by the Author

The first time I realized the sky wasn't as innocent as it looked, I was standing outside a forward operating base under a morning that seemed too perfect for war. The air was still, the horizon quiet. But above us, unseen and unheard, something was always watching.

That memory stayed with me long after the uniform faded into the evening. Years later, as technology blurred the line between being a soldier and a civilian, that same stillness began to feel different, not peaceful, but vulnerable. We had entered a new kind of battle space, one without front lines, flags, or even declarations of war.

The tension between admiration and apprehension, faith and function, human choice and machine autonomy defines this book. The narrative explores the consequences when progress outruns principle, policy fails to keep pace with technological propulsion, and innovation neglects those it should benefit.

I wrote this as both a Command Sergeant Major and a father, as someone who has seen how easily vigilance can slip into complacency, and how technology can make us forget the cost of human judgment.

I hope that every reader, soldier, student, or citizen walks away from these pages with a renewed sense of ownership over the sky we share. Because leadership doesn't begin with a rank or a title, it starts with awareness.

As I've said before, the future isn't waiting to be discovered; it's waiting to be decided. And the sky? It's not empty. It's a reflection of who we are and who we still have time to become. CSM Sheldon A. Watson (USA)

Chapter 1– Awakening the American Sky, The Unseen Threat

Section 1: A New Era of Airspace

It began, as many modern headaches do, with a viral video. A teenager in Kansas, armed with a $200 drone and a questionable sense of patriotism, decided to celebrate the Fourth of July by strapping fireworks to his airborne toy. The result? A million views in two days, a scorched shed, and a fire department callout. The drone is lost. The neighborhood sighed. And the lesson?

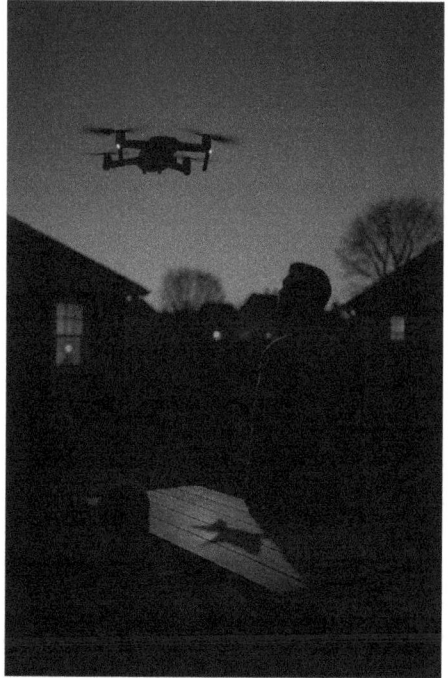

Apparently, none. The clip was hilarious. The implications were not. That teenager inadvertently demonstrated how a consumer-grade gadget, in the wrong hands, could spark chaos worth millions of dollars.

Marshall McLuhan once said, "We shape our tools, and thereafter our tools shape us."

He wasn't wrong. He just didn't anticipate that those tools would fly one day.

Once upon a time, anything that moved through the sky triggered NORAD's attention. Today, it might just be your neighbor's new Christmas toy, hovering over the backyard, filming your barbecue, or drifting aimlessly because someone got bored after dinner. In less than two decades, flight has shifted from the realm of fighter pilots and astronauts to teenagers, farmers, and vloggers. The sky, once a symbol of national strength and strategic control, has become a shared commons of curiosity and commerce. When the Wright brothers lifted off at Kitty Hawk, the sky was infinite. When Lindbergh crossed the Atlantic, it stood as a symbol of courage and determination. When NASA planted a flag on the moon, it embodied possibility. Now, that same sky signals vulnerability.

Technology democratized flight, but it also decentralized responsibility. Every new drone owner inherits a sliver of national airspace and, whether they know it or not, a slice of national security. The drone didn't sneak into history; it arrived with fanfare, sleek packaging, and influencers showing off cinematic sunsets. By the time regulators began drafting rules, millions had already taken to the air. What started as a tool for aerial photography morphed into a revolution in logistics, agriculture, construction, and emergency response. Drones inspect bridges, monitor crops, deliver medicine, and help firefighters see through smoke. They also peek through windows, interfere with aircraft, and smuggle contraband across borders. That's the paradox: the same innovation that saves lives can also end them.

Airpower used to be measured by altitude and payload. Now, it's measured by accessibility. Anyone with a credit card and Wi-Fi can own what once required a defense contract and a runway. During my first deployment, unmanned systems were mysterious, expensive, and restricted. They demanded operators, analysts, and clearances. Today, that same capability fits in a backpack and requires no clearance. The barrier between military and civilian technology has collapsed. The same sensors we used for reconnaissance are now sold in retail stores. The same aerial perspective that gave commanders strategic advantage now gives YouTubers cinematic flair. It's tempting to call that progress, and in many ways, it is. But when technology spreads faster than education, we trade understanding for convenience. The result? Millions of micro-pilots, few of them trained, all of them sharing one sky.

Americans love freedom. We cherish the right to move, build, explore, and innovate. But freedom without boundaries quickly becomes fragile. In aviation's early days, pilots operated under mutual respect: eyes on the horizon, rules of the air, shared responsibility. Drones disrupted that balance. The sad truth is that today's pilot might not even be outdoors; they could be behind a screen, miles away. Traditional aviation relied on visibility and accountability. Drone aviation often relies on anonymity. And that anonymity is precisely what adversaries exploit. To the naked eye, there's no difference between a drone delivering dog food and one delivering danger. Both weigh under ten pounds, both use GPS, and both can be bought online. The intent, not the technology, defines the threat. And intent is invisible until it acts.

Washington's response to the drone boom? Bureaucratic whiplash. Agencies scrambled to catch up. The FAA had to redefine "aircraft" in the age of quadcopters. DHS wondered whether every hovering object posed a security risk. The Department of Defense debated when it could legally engage in such an action.

Meanwhile, the drones kept flying. Regulations that once took years to draft were obsolete upon release. Rules designed for airplanes couldn't keep pace with pocket-sized pilots. Cities banned drones near airports; hobbyists protested. Civil liberties groups raised privacy alarms; security agencies raised threat levels. Somewhere between freedom and fear, policy stalled, and the sky kept getting busier.

Part of the challenge is cultural. Americans see drones as entertainment, not infrastructure. We don't associate a buzzing quadcopter with national security; we associate it with halftime shows and backyard stunts. We scroll past viral drone footage without realizing we're watching the future of reconnaissance. We laugh at delivery mishaps without recognizing the logistics revolution underway. And when a drone disrupts an airport, we treat it as curiosity rather than a warning. The problem isn't just that people don't understand drones; it's that they don't think they need to. But the next time your flight is delayed due to a "UAS incident," remember that's not a rare fluke. That's the sound of a new normal knocking on the hangar door.

Drones occupy a literal and metaphorical blind spot. They operate below traditional radar, beyond the line of sight, and often beneath public consciousness. To the average citizen, airspace is an abstract concept. We think of it only when we fly commercially or spot a jet contrail overhead. Yet between 100 and 400 feet, the space between treetops and skyscrapers, a quiet revolution unfolds daily. This invisible frontier is now where the most critical intersections occur: commerce meets security, innovation meets oversight, and freedom meets accountability. We're not just living under the same sky; we're sharing it with machines that see, record, and sometimes decide.

It's easy to think drone security is someone else's problem, the FAA's, the military's, law enforcement's. But airspace, like democracy, is a shared resource. Every citizen benefits from it. Every citizen has a stake in protecting it. The same vigilance we teach for neighborhood safety, "See something, say something," now extends upward. If a neighbor's drone behaves strangely or flies where it shouldn't, awareness isn't paranoia; it's prudence. Because in this new era of airspace, inaction is a luxury we can no longer afford.

As we marvel at the convenience drones bring, faster deliveries, sweeping videos, and effortless inspections, we must confront the darker side of that accessibility. The same design that makes them helpful makes them weaponizable. The line between toy and tactic has all but vanished. And that brings us to the next section, where we explore how a tool meant to entertain became a tool that can intimidate: *"From Toys to Tactics, The Dual-Use Dilemma."*

Section 2: From Toys to Tactics: The Dual-Use Dilemma

"Technology is neither good nor bad; nor is it neutral." Melvin Kranzberg's first law of technology isn't just clever, it's prophetic.

Especially when applied to drones, the poster child of modern dual-use dilemmas. They arrived in glossy boxes, promising creativity and convenience: "Capture your adventures," "See the world from above." The ads looked more like GoPro commercials than national security briefings. Parents bought them for Christmas. Teenagers filmed skateboard tricks. Real estate agents finally got their cinematic property shots. No one imagined that, just a few years later, similar models would be dropping grenades through open hatches or mapping troop movements in real time. That's the illusion of innocence. The same drone that documents your wedding can also scout your water-treatment plant. The threat isn't in the device; it's in the intent.

Historically, governments led technology. Drones flipped that script. Consumers sprinted ahead while agencies fumbled to define what they were even dealing with. By 2015, the FAA projected that there would be 1 million drones in U.S. airspace. It is underestimated by half. Meanwhile, tech companies were releasing smarter, faster, more autonomous models every six months. Innovation sprinted; policy stretched. The result? A vacuum. And nature, especially human nature, abhors a vacuum. That vacuum got filled by curiosity, commerce, and eventually, cunning.

Consider the drone's journey. A kid in Arizona uses one to photograph the desert at sunset. A cartel in Mexico uses one to deliver fentanyl across the border. A militia in the Middle East uses one to drop mortar rounds. A protester in Europe uses one to live stream a demonstration. Same hardware, four motives. That's dual use in action, a tool equally capable of good and harm depending on whose hands control it. During the Iraq and Afghanistan conflicts, U.S. forces first encountered this shift when commercial quadcopters began appearing over patrol routes. At first, troops assumed they were journalists or curious civilians. Then the pattern changed: drones hovered longer, followed convoys, or vanished moments before indirect fire landed. The battlefield had evolved. No one sent a memo.

In warfare, economics matters. When a $300 drone disables a $5 million vehicle, that's not just a tactical loss, it's strategic shock. Ukraine made that painfully clear. Footage from the conflict shows low-cost drones dropping munitions into tank hatches, disabling artillery, and targeting radar systems. Operation "Spider Web," as some analysts dubbed the swarm tactics, turned consumer tech into battlefield dominance. Every defense planner in the world took notice. The lesson was blunt: the future of warfare had gone retail. Back home, that should have triggered a nationwide conversation about infrastructure security. Instead, it triggered hashtags and highlight reels.

Here's where it gets complicated. The same companies designing drones for cinematography and farming are, by extension, shaping the future of surveillance and warfare. When engineers build more intelligent navigation systems, they're not thinking about adversaries evading radar. When developers improve battery life, they're not imagining someone weaponizing endurance. Yet those improvements serve both. It's the paradox of progress: every step toward convenience casts a shadow of consequence. This isn't an indictment of innovation; it's a call for awareness. A drone that saves a farmer's season shouldn't endanger a soldier's life. But without ethical foresight, we risk both.

Knowledge used to be a strategic advantage. Now it's a search bar away. Type "how to modify a drone for payload" and the algorithm will cheerfully serve up step-by-step tutorials. What once required military training can now be learned between coffee breaks. Online forums share 3D-printed mounts, signal-boosting tricks, and evasion tactics. It's democratized warfare in the worst possible sense. For the average citizen, that raises a straightforward question: if anyone can learn it, who's ensuring they don't use it?

Defending against drones isn't as simple as shooting them down. Legally, it's nearly impossible for anyone other than specific federal entities to interfere with aircraft, and yes, under the law, a drone is an aircraft. So, when a suspicious drone appears over a stadium or refinery, local law enforcement often must, yes, you guessed it, sit and watch. They can't jam or intercept the signal without risking federal violations. The bureaucracy, designed for a 747 aircraft, now governs devices that fit in a glove compartment. Meanwhile, the adversary has no such limitations. That's the asymmetry: offense costs little and obeys no rules; defense costs millions and moves through red tape.

Part of what sustains the dual-use problem is our collective detachment. Drones seem digital, as if they exist somewhere in "the cloud," rather than in our neighborhoods. But make no mistake: every drone is physical, with a motor, a payload, and a purpose. It's as real as the operator's intent. We've become conditioned to view risk as something online, rather than airborne. Yet the same technology that lets you stream aerial footage can also stream live reconnaissance for those plotting harm. It's not paranoia, it's physics. What goes up can come down on the wrong target.

Add to this the uncomfortable reality that much of America's drone hardware and software originates overseas. That means potential backdoors in firmware, unverified data routing, and foreign dependency for replacement parts. Imagine if 80 percent of our police radios came from a geopolitical rival, we'd call that a vulnerability. Yet for drones, that's normal. Security analysts warn that until domestic production scales up; our skies remain at least partially outsourced. The question isn't just who's flying, it's who built what they're flying.

Artificial Intelligence adds another layer. The newest consumer drones can track faces, follow vehicles, avoid obstacles, and return home autonomously. Now imagine that same autonomy programmed for malicious ends. The next generation won't need human pilots or coordinates. Swarms will cooperate, learn, and adapt to their environment. It sounds futuristic, but it's already in prototype. That's why awareness today matters. You don't prepare for tomorrow's treatment after it arrives; you prepare while it's still optional.

Despite all the circuitry and code, the most unpredictable component remains human behavior. We innovate faster than we educate. We crave capability before comprehension. That's the heart of the dual-use dilemma: we're brilliant at invention and terrible at restraint. Every generation wrestles with its new power: the printing press, the atom, the Internet. Ours just happens to fly.

And so, the dual-use nature of drones leaves us in a moral and strategic gray zone. What began as curiosity is now capability. What was entertainment is now escalation. In the next section, *"When Curiosity Becomes a Weapon,"* we'll examine how that gray zone has already produced real-world consequences: airports shut down, power grids probed, and communities caught off guard. Because the question is no longer whether drones will be misused, it's how ready we'll be when they are.

Section 3: When Curiosity Becomes a Weapon

"Every tool is a weapon if you hold it right," Ani DiFranco once quipped.

She wasn't talking about drones, but she might as well have been. Every technological revolution begins with curiosity, and if history is any guide, it ends with regret. Drones followed that arc with alarming precision. What started as a marvel of innovation quickly morphed into a buzzing, airborne Pandora's box, compact, accessible, and capable of consequences far beyond its size.

The first major wake-up call didn't come from a battlefield. It came from Gatwick Airport, London's second largest, in December 2018. For nearly three days, the airport went silent. Flights were grounded, travelers stranded, and security forces scrambled. The culprit? Repeated drone sightings near the runway. Not missiles. Not sabotage. Just drones. More than 140,000 passengers were affected, 1,000 flights were canceled, and the economic impact soared into the tens of millions. No bombs dropped. No shots fired. Just a consumer-grade quadcopter bringing one of Europe's busiest transportation hubs to its knees. To this day, no one knows who flew them. That incident proved a chilling point: you don't need to destroy something to disrupt it.

After Gatwick, airports worldwide quietly rethought their security posture. Detection systems were installed. New "no-fly zones" were mapped. Federal agencies dusted off contingency plans. But the truth was already airborne: prevention had failed before policy even began. For the first time, the world saw that a hobbyist tool could inflict the same level of disruption as a coordinated terror attack, without any formal declaration, organization, or motive. Thus, a new category of threat emerged: the weapon of inconvenience. Not every attacker needs to destroy; some just need to delay, confuse, or cause panic. And drones, with their speed, accessibility, and plausible deniability, are tailor-made for it.

Soon, curiosity expanded beyond airports. Security teams began spotting drones near critical infrastructure, including power substations, refineries, and communications towers. In 2022, the FBI and DHS investigated multiple drone incursions over energy facilities in the American Midwest. Some carried cameras. Others carried nothing visible at all. The operators? Never identified. Were they hobbyists? Activists? Foreign proxies? No one could say for sure. And that uncertainty is the weapon. Unlike a missile, a drone doesn't announce intent. It hovers, observes, waits. While law enforcement sorts out jurisdictional boundaries, the operator, whomever they are, gathers information in real-time. Every pass of a camera-equipped drone over a substation, police facility, or port is a potential reconnaissance mission. In this era, data is the new ammunition.

While policymakers debated drone ethics, criminal networks got creative. Along the U.S.–Mexico border, drones became the contraband courier of choice. DEA reports estimate hundreds of drones cross the border each month, carrying narcotics, cash, or surveillance gear. They fly low, fast, and often at night, small enough to evade radar, cheap enough to replace. Smugglers adapted faster than regulators. A drone lost to wind or interference isn't a tragedy; it's a tax write-off. This evolution from curiosity to criminality underscores a brutal truth: the barrier to entry for aerial misconduct has been largely eliminated. It's not about sophistication, it's about creativity. And criminals have plenty of that.

Even domestic actors aren't immune to temptation. In 2021, a man in Pennsylvania was arrested for using drones to drop contraband into a state prison. His equipment? Two consumer quadcopters and a homemade release mechanism. No specialized background. No military training. Just YouTube, spare time, and the wrong intentions. Similar incidents have occurred at prisons nationwide, small, improvised deliveries carrying everything from drugs to cell phones. Each event reinforces the same lesson: we've handed everyone an air force, and we're shocked they're using it.

Strategists call it the gray zone, the space between peace and war, where mischief doesn't qualify as aggression but still causes harm. Drones thrive in that ambiguity. A drone hovering over a refinery isn't an attack, but it's not harmless either. One flying near a military base may be testing radar response times. Another outside a politician's home might be scouting security patterns. Yet none of these actions neatly fit existing laws. Until it drops something or crashes intentionally, a drone remains protected by ambiguity. That's the challenge: it exploits our moral hesitation and legal red tape.

Fear fatigue is real. After years of headlines about data breaches, cyberattacks, and mass shootings, the public grows numb. Drones add another layer of invisible anxiety. Most Americans don't want to imagine that the same drone taking wedding photos could also be testing power-plant vulnerabilities. But denial doesn't prevent danger; it just delays recognition. Communities have reported drones following school buses, hovering near hospitals, and flying through neighborhoods at night. In most cases, they're dismissed as "harmless curiosity." But as history has taught us, complacency is rarely harmless.

The real power of drones lies not in the damage they cause, but in how easily they cross boundaries, both physical and psychological. Fences, gates, and walls mean little to a device that flies over them. Traditional defenses, built for human intruders, are powerless against machines operated from a couch. This creates a new category of trespass, one that can't be stopped by locks or lights, only by awareness and collaboration. The irony is thick: the technology that gives us unprecedented visibility from above has also exposed just how blind we are below.

As drones continue to evolve, smaller, quieter, and more autonomous, the line between innocent flights and hostile reconnaissance will blur even further. Artificial intelligence will soon enable drones to operate independently, identify targets, and evade detection without human intervention. The gap between hobbyists and hostile will completely vanish. That's why this isn't just a policy discussion, it's a public awakening. If people don't understand the risks, they can't recognize the warning signs. And if they can't recognize them, no amount of military readiness will matter. The fight for the American sky isn't only about technology; it's about teaching the nation to look up.

In the next section, *"Ignorance Isn't Innocence,"* we'll unpack the cultural and psychological roots of our national blind spot. Why do people ignore what's above them? Why does the average citizen assume the government has it handled? And how does that complacency quietly empower those who mean harm? Because awareness without action is just curiosity, and as we've seen, curiosity can be weaponized.

Section 4: Ignorance Isn't Innocence

"The greatest danger to our future is apathy," Jane Goodall warned.

She wasn't talking about drones, but she might as well have been. In America, ignorance rarely stems from stupidity; it's rooted in comfort. We're a nation built on the quiet assumption that someone, somewhere, is handling the hard stuff. When a plane flies overhead, we trust air traffic control has it. When an emergency strikes, we assume first responders are on the way. When a strange object appears in the sky, most people shrug and believe it's someone else's problem. We prefer our skies to be quiet and our consciences to be clear.

The trouble is that mindset no longer works. The systems we trust to manage danger were built for a different kind of sky, one where every aircraft had a flight plan and every threat wore a uniform. Today's sky is fragmented, anonymous, and shared. The new frontier of risk floats right above our homes, yet most people still treat it like science fiction.

We've always measured danger by proximity, floods, fires, storms, things we can see and hear. Drones violate that instinct. They're quiet, small, and often invisible against a bright sky. You could walk your dog under an active drone reconnaissance operation and never know it. This sensory gap creates a psychological one. If it doesn't feel like a threat, it doesn't register as one. It's the same blindness that led early internet users to share personal data freely: "What's the harm?" Only later did they learn that the digital space they ignored had rules, predators, and consequences. The sky is now experiencing its own version of that naïveté. We assume safety simply because danger hasn't yet presented itself.

After decades of high-tech warfare and 24-hour news cycles, Americans have developed a dangerous reflex: we outsource vigilance. We assume defense is the Pentagon's problem, surveillance is the FBI's concern, and someone with a badge is always watching the watchers. But no agency can monitor every inch of our sky, not above 300 million rooftops, not over every substation, stadium, or schoolyard. This isn't neglect, it's physics. The coverage area is too vast, and the technology is too cheap and mobile. The result? A gap so wide that the average citizen could fly a drone from their backyard over sensitive infrastructure without anyone knowing, or caring, until it's too late. In this environment, ignorance isn't neutral. It's permission.

Mainstream media hasn't helped. Drones are either glorified or vilified, depending on the headline. When a company announces a new delivery drone, it's hailed as futuristic innovation. When one interferes with firefighting aircraft, it's treated as a quirky outlier, wedged between weather updates and celebrity gossip. But both stories belong to the same narrative: the rapid normalization of ungoverned airspace. By treating drones as novelties rather than necessities, the media shaped public perception into one of indifference. They're seen as gadgets, not gateways; toys, not tools. And when the public stops paying attention, accountability tends to fade.

Here's the irony: the safer we feel, the less secure we often are. Technology has given us instant information, but it's also numbed our instinct to act on it. We scroll through global crises, earthquakes, wildfires, invasions, and think, "That's terrible," then swipe to the next post. Drones fall into that same emotional blind spot. People see footage of attacks overseas and say, "That's tragic... over there." The problem is that "over there" is only one firmware update away from "right here." Familiarity breeds detachment. And complacency is the oxygen of vulnerability.

Somewhere between Hollywood thrillers and Homeland Security briefings, Americans developed a belief that the government sees everything: satellites, radars, sensors, the Big Eye in the Sky. That illusion breeds apathy. If "they" have it covered, why should "we" worry? Here's the uncomfortable truth: most drones flying below 400 feet go completely undetected by traditional radar systems. They're too small, too low, and too quiet. The federal response network isn't omniscient; it's overstretched. Even when incidents are reported, jurisdictional confusion reigns. The FAA handles flight safety. DHS handles threats. The FBI handles criminal acts. The DoD steps in if national defense is implicated. By the time everyone figures out whose sandbox it is, the drone is long gone. The sky, once unified under control, is now a patchwork of policies, and bad actors exploit the seams.

Pilots have a phrase: "Get-there-itis." It's the mindset that leads to accidents when a pilot becomes so focused on reaching the destination that they ignore danger signs along the way. As a nation, we've developed "normalcy-itis." We're so used to our daily rhythm that we stop questioning the unfamiliar hovering around it. We've normalized drones over stadiums, neighborhoods, and parades. We treat them as background noise. But normalization without understanding is complacency in disguise. And complacency is a luxury no free nation can afford.

If ignorance fuels vulnerability, education is the antidote. Yet drone literacy, understanding what they are, how they work, and why they matter, is almost nonexistent in schools or public safety campaigns. We teach children about cyberbullying, but not about online privacy. We hold fire drills but not drone awareness sessions. The next generation will inherit skies far more crowded and complex than ours, and we're sending them in blindfolded. Awareness isn't paranoia. It's preparedness. The sooner the public learns that distinction, the sooner we close the gap between curiosity and catastrophe.

If there's one thing the military teaches you, it's to find humor in hard truths. So here's one: Americans are terrified of self-driving cars but completely fine with flying robots. One malfunctions in traffic and it's national outrage; the other drops from the sky and we call it "an unfortunate glitch." We've built an entire culture around reacting to what we can see, while the real danger hums above us like a mosquito we refuse to swat. A little irony never hurts, as long as it wakes people up.

Ignorance is a shield until it becomes a shroud. Eventually, the things we choose not to see become the things that define us. If the last century taught us anything, it's that national security isn't built solely on weapons or walls; it's built on awareness. From neighborhood watches to cyber hygiene, safety starts where knowledge lives. Every American who learns to distinguish between a hobbyist drone and a suspicious flight adds one more set of eyes to the nation's defense, and every local leader who invites law enforcement to brief their community chips away at the blind spot. Awareness doesn't just protect, it empowers.

In the next and final section, *"A Shared Sky: The Public's Stake in Aerial Security,"* we'll explore how that empowerment turns into action because knowing the threat isn't enough. The objective measure of a nation isn't how much it fears, but how much it responds. The sky belongs to all of us. And in the age of drones, so does the responsibility.

Section 5: A Shared Sky: The Public's Stake in Aerial Security

"Ask not what your country can do for you, ask what you can do for your country." JFK.

JFK's words still echo, but in the age of drones, they carry a new altitude. For most of our history, Americans looked up and saw ownership in symbols: the eagle, the flag, the contrail of a jet streaking toward purpose. The sky meant freedom. Now it also means responsibility.

No single agency, corporation, or military command can protect the nation's airspace alone. There are too many miles, too many machines, and too many motives. Every citizen who benefits from the sky's convenience shares a small part of its defense. If that sounds daunting, remember this: the same technology that threatens us also empowers us. Drones don't just create vulnerability; they create visibility. They remind us that vigilance isn't only vertical; it's communal.

Security used to be something done for the people. In this era, it must be done with them. Imagine if the same energy poured into neighborhood-watch programs were to extend upward. The principle is the same: eyes, awareness, accountability. When a resident reports a strange drone near a refinery or stadium, they're not being paranoid; they're practicing patriotism. Local law enforcement already depends on public input for ground-level crimes; now it needs the same cooperation for threats in the air. A simple phone call can trigger pattern analysis that connects dots across jurisdictions. The goal isn't to turn citizens into spies, it's to turn apathy into attention.

The biggest mistake we can make is to frame drone awareness as fear. Fear paralyzes; partnership mobilizes. Public-private partnerships are already redefining resilience in cybersecurity, energy, and emergency management. The same model can secure our skies. Utility companies can share sensor data with local authorities. Universities can run drone literacy programs. Even hobbyist clubs can play a role in education and self-regulation. It's not about creating a police state, it's about creating a cooperative state. When the public understands the "why," compliance follows naturally. People resist control but respond to clarity. Give them the facts, the stakes, and the channels to act, and they'll surprise you with how much they care.

Despite headlines of division, America still thrives on shared purpose. Every disaster, every crisis, every challenge proves the same point: when things go wrong, ordinary people step up. Hurricane survivors formed volunteer flotillas. Neighborhoods organized food drives during the pandemic. Veterans mentored youth through service programs. The pattern is consistent: when the call is clear, Americans answer. Drone awareness is no different. You don't need clearance or a uniform to contribute; you just need curiosity, courage, and a sense of community. The first line of defense for the homeland has always been its people.

If ignorance fuels vulnerability, education builds armor. Drone awareness doesn't have to be complex; it just has to be consistent. Elementary schools can include lessons on responsible tech use. High schools can discuss the ethics of privacy and innovation. Community colleges can partner with local agencies to train drone operators in safety and situational awareness. Imagine a future where every graduating senior understands not only how to operate a drone but also how to recognize its misuse. That's not paranoia, it's preparedness. Education doesn't just prevent accidents; it deters intent. When people know they're being watched responsibly, bad actors think twice.

With every new domain of technology comes an ethical frontier. We must teach not just how to fly, but why and when it's right to do so.

Freedom of flight, like freedom of speech, comes with boundaries. The First Amendment doesn't protect threats; neither should the sky. Ethics bridges the gap between law and conscience. It's what keeps innovation human. When hobbyists respect privacy, corporations respect regulations, and governments respect transparency, the air above becomes safer for everyone beneath it.

National defense is a mosaic of local actions. One alert citizen reporting a drone near a water-treatment facility can trigger an investigation that prevents a cascading failure. Programs like "If You See Something, Say Something" need a 21st-century upgrade: "If You See It, Report It, Even If It Flies." Technology can aid that mission. Mobile apps could allow geotagged drone reports. AI systems could analyze trends and identify hotspots of suspicious activity. This isn't science fiction; it's civic modernization. Every town hall that adds a two-minute drone briefing, every sheriff's office that posts awareness flyers, every newsroom that treats incursions as serious news rather than novelty, each strengthens the invisible network that keeps our nation secure.

Let's be honest: the idea of crowdsourcing homeland defense sounds like something between a reality show and a sci-fi movie. "America's Got Vigilance." But humor has a purpose here. It reminds us that ordinary people, when informed and inspired, can accomplish extraordinary things. We don't need paranoia, we need participation. And participation works best when people believe they matter, because they do. No general, no sensor, no satellite can replace the eyes of a community that cares.

A free nation must constantly balance liberty with security. The conversation about drones is, in essence, a conversation about trust, trust between citizens and government, between innovation and oversight, and between convenience and conscience. True freedom isn't doing whatever we want; it's ensuring we can keep doing it safely. That means making small sacrifices for the greater good: following flight rules, reporting suspicious activity, and resisting the temptation to dismiss the sky as "someone else's problem." In the end, democracy isn't defended in Washington; it's defended in neighborhoods.

The sky may no longer be empty, but it is still ours. The responsibility to protect it isn't confined to cockpits or command centers; it lives in classrooms, living rooms, and backyards. Every American who chooses awareness over apathy becomes part of the shield that guards this nation. That's not exaggeration, that's citizenship. As technology accelerates and threats evolve, our greatest weapon remains unchanged: an informed, engaged public that refuses to look away.

In the next chapter, *"Dual-Use Technology, Innovation's Double-Edged Sword,"* we'll look deeper at the paradox that drives this entire challenge, how the same ingenuity that built a safer, smarter, more connected world also opened the door for exploitation, and how we can reconcile innovation with accountability without sacrificing either. Because in the end, this isn't a story about drones. It's a story about us, a nation learning to protect what it loves before it loses sight of it.

Chapter 2 – Dual-Use Technology, Innovation's Double-Edged Sword

Section 1: The Paradox of Progress

It began, as modern miracles often do, with an inconvenience. A suburban dad in Colorado ordered a smart coffee maker online. Two days later, he stood in his driveway watching a delivery drone descend with his morning salvation. Neighbors peeked out from behind curtains. The kids cheered. The dog lost its mind. The drone touched down, released the package, and zipped away, almost. Halfway across the street, it clipped a pine tree, spiraled, and smacked into a mailbox. By evening, the video had a million views under the caption, "When Skynet Delivers Late." America laughed. But beneath that laughter was a familiar tension, the one between invention and control. From railroads to nuclear reactors, every great leap forward has been accompanied by a quiet step backward.

As John W. Gardner put it, "We are continually faced with great opportunities which are brilliantly disguised as insoluble problems."

America doesn't invent because it's easy; we invent because we're wired that way. It's cultural muscle memory; no problem is unsolvable if you throw enough creativity and caffeine at it. From the cotton gin to the microchip, innovation has defined our national identity. But every breakthrough arrives with an uninvited twin: consequence. Eli Whitney's cotton gin revolutionized the industry and inadvertently deepened slavery. The automobile gave freedom and traffic deaths. The Internet gave connection, and chaos. The pattern repeats: progress always asks for payment. Drones are simply the next entry in that long ledger, a symbol of ingenuity and a reminder that tools don't choose how they're used; people do.

For most of human history, the sky was off-limits. It belonged to birds, gods, and dreamers. Then the Wright brothers lifted off, and suddenly, we were tenants in the heavens. Each century since has lowered the rent. Airplanes gave way to satellites, which in turn gave way to rockets, which gave way to reusable boosters. What once required a nation-state can now be achieved by a teenager with a soldering kit. The line between imagination and implementation has vanished. Inventions once confined to DARPA labs are now weekend projects on YouTube. Flight, once one of humanity's oldest fascinations, has become a personal hobby. And with that shift, the sky stopped being sacred and

27

began to be crowded. Progress, unregulated, rarely pauses to ask whether it should, only whether it can.

There's an arrogance that comes with being a technological superpower. We assume progress equals improvement, that faster is always better, smarter is always safer, and smaller is always cheaper. But history disagrees. The same circuits that guide a drone's flight can guide a missile. The same AI that tracks wildlife can track people. The same GPS that finds your pizza shop can find a convoy. In the race to innovate, America often forgets the finish line isn't speed, it's stewardship. Innovation without foresight is like flying without a map: exhilarating right up until the crash.

During World War II, scientists in New Mexico cracked the atom, an achievement so profound it redefined power, politics, and philosophy in one blinding flash. Yet even as the first mushroom cloud rose, Robert Oppenheimer whispered, "Now I am become Death, the destroyer of worlds." It wasn't remorse for science; it was recognition of duality. Creation and destruction often wear the same lab coat. Today, that same duality plays out in boardrooms and garages across America. Engineers design machines to make life easier, safer, and more connected. And someone, somewhere, figures out how to turn those same machines into dangers. We no longer invent in isolation; we invent in a global echo chamber, where every blueprint is both an opportunity and a liability.

"Move fast and break things." That was Silicon Valley's motto, right up until the things that broke weren't just software, but systems. Technology companies once sold tools. Now they sell trust. And trust is fragile. The pursuit of convenience often skips the step of consequence. For years, the drone industry has raced to outdo itself, with longer flight times, higher payloads, and increased autonomy. Few paused to ask what would happen when those same improvements fell into the hands of hostile forces. When innovation outpaces introspection, disruption can become destruction.

It's not just policy that drives American innovation; it's pride. Our collective ego is tied to progress. We built skyscrapers that scraped the heavens, railroads that stitched together continents, and networks that digitized reality. To slow down feels un-American. But somewhere between Edison's light bulb and the latest drone prototype, innovation shifted from public mission to private race. The focus moved from nation-building to market share. We stopped asking, "What does this do for us?" and started asking, "How fast can we sell it?" The unintended result: a society more connected, yet less secure.

Progress has a speed limit, it's called understanding. Every time we exceed it, we lose control. The Wright brothers studied airflow for years before achieving their first flight, lasting 12 seconds, at Kitty Hawk. Today's startups push new hardware to market every six months with minimal testing. The gap between idea and impact has evaporated. The same impatience that fuels creativity also breeds complacency. We assume someone else will handle the safety checks, someone else will write the rules. But "someone else" is an endangered species in the age of autonomy. When everyone innovates, no one takes ownership.

The story of human progress can be summed up in six words: We built it. It built us. The printing press gave us knowledge and propaganda. The radio gave us music and demagogues. The Internet gave us truth and disinformation. Now, drones provide us with freedom and exposure. They deliver medicine to disaster zones and surveillance to dictatorships. They monitor crops and monitor citizens. Every tool carries the DNA of its maker: ambition, imagination, and imperfection. The drone, in all its efficiency, is simply a mirror reflecting our own contradictions.

Every generation faces a defining question about its inventions: Will they serve us, or will we serve them? Ours is no exception. We've created a technology so powerful, so versatile, that it forces us to confront the limits of our own discipline. Drones are not evil; they are indifferent. It's the user who decides whether the sky brings aid or destruction. Thus, the challenge before us isn't to stop innovation, it's to civilize it. To reintroduce moral gravity into a world obsessed with flight.

In the next section, "*Innovation Without Instruction: When Capability Outruns Conscience*," we'll explore how the rapid pace of technological growth outstrips our ability to regulate or even comprehend it. Because in the end, progress without principle isn't leadership, it's drift. And in the age of drones, drift can be deadly.

Section 2: Innovation Without Instruction: When Capability Outruns Conscience

"Our technology has surpassed our humanity." Albert Einstein (quoted during the dawn of the nuclear age).

He wasn't lamenting progress; he was cautioning against its pace. In America, "new" is a religion. We bless every gadget, every app, every breakthrough as proof that we're still ahead. But in our race to out-invent the world, we've stopped asking a simple question: Should we? The problem isn't that innovation is dangerous; it's that it's undisciplined. When the guiding principle is "move fast and disrupt," no one pauses long enough to read the fine print on disruption. Drones are the perfect example. We built them before we built boundaries. Then, once the skies were full, we wondered why it felt crowded.

The average person today carries more processing power in their pocket than NASA used to land the Apollo 11 mission. Yet we still treat that power like a toy. Technological literacy hasn't kept pace with technological access. We often use things long before we fully understand their implications. It's the equivalent of handing a teenager the keys to a sports car and hoping they'll "figure it out on the highway." Every advance creates two timelines: one for progress and one for understanding. The first moves at the speed of profit; the second crawls at the pace of reflection. When capability outpaces conscience, history always sends the bill later.

We've seen this movie before. During the Cold War, superpowers raced to stockpile weapons they barely understood. Scientists warned about fallout; politicians wanted deterrence. The logic was simple: if we can, we must. Fast-forward to today, and the same reasoning governs drones, AI, and biotech. The arenas have changed, but the psychology remains: power is addictive, and foresight is inconvenient. Only now, the arms race isn't between nations, it's between corporations, coders, and consumers.

In the tech world, speed isn't a metric; it's a virtue. Launch now, patch later. That mindset works for video games. It doesn't work for machines that fly, sense, and record. A single coding error in a social media app could lead to embarrassment. A coding error in an autonomous drone could result in fatal consequences. But the industry thrives on iteration, not introspection. The result is what I call "ethics lag", the time gap between what technology can do and what society should allow it to do. The longer that gap stays open, the more risk seeps through.

Inventions often start with a necessity: "How do we grow more food?" "How do we move faster?" Now they start with curiosity and venture funding: "What if we could...?" The pursuit of the "next big thing" rarely includes a plan for "the next bad thing." Consider facial-recognition technology. It began as a security tool, became a marketing gimmick, and now fuels debates on privacy and bias. Drones are following the same trajectory, from defense asset to delivery service to surveillance concern. The motive isn't malice, it's momentum. Because the truth is, progress doesn't pause for moral check-ins.

Here's the cruel irony: every generation thinks it's the first to deal with unintended consequences. Yet every innovation, from gunpowder to the Internet, has taught the same lesson: capability without context leads to chaos. We just keep graduating from the same course without learning the material. In drone technology, that lesson is glaring. Engineers perfect autonomous navigation but rarely study its ethics. Marketers pitch speed and convenience but sidestep security. Regulators try to write rules for technology they barely comprehend. It's not a lack of intelligence; it's a lack of integration. We silo science from ethics, policy from practice, and then act surprised when the results collide.

Schools teach coding before they teach consequences. We celebrate STEM, but we often forget to add the "H" for humanity. Imagine if every engineering program required a semester of public-policy awareness or moral philosophy. Imagine if every drone manufacturer had to conduct a social-impact review before releasing a product. These ideas sound idealistic until you realize the alternative is what we have now: brilliant minds designing in moral isolation. Technology isn't born evil. It's raised that way, in environments where "can we?" always beats "should we?"

Meanwhile, the government struggles to keep up. By the time an agency drafts a rule, the technology it targets has already evolved. It's like chasing a jet with a bicycle. In 2016, when the FAA released its first comprehensive drone regulations, the industry was already two generations ahead. By the time Congress debates one ethical dilemma, three new use cases appear. We're not just behind, we're structurally behind. Bureaucracy was built for permanence. Technology thrives on change. Unless adaptive governance becomes the norm, we'll continue to legislate yesterday's problems tomorrow.

People love drones because they find them fascinating. They turn ordinary citizens into pilots and explorers. But fascination is a double agent; it distracts from risk. Every time a viral video shows a drone rescue or a sunset flyover, it reinforces the illusion that these machines are inherently good. The public sees the spectacle, not the shadow. And when admiration replaces accountability, innovation becomes entertainment rather than a responsibility. We don't need to fear the technology; we need to mature alongside it.

Humanity has always raced ahead of its reflection. We invented fire before we had fire codes, the car before seatbelts, and the Internet before the delete button. So, the question isn't whether drones will change the world; they already have. The question is whether we'll guide that change with foresight or clean up after it with regret. Ethics isn't the brake pedal on innovation; it's the steering wheel. Without it, progress doesn't stop; it just crashes somewhere you didn't intend to go.

In the next section, *"Made for Commerce, Used for Conflict,"* we'll follow the drone from the factory floor to the battlefield and examine how the global market blurred the line between civilian and military technology. Because when capability outruns conscience, someone will always find a way to profit from the fallout.

Section 3: Made for Commerce, Used for Conflict

"The distinction between the civilian and the military is increasingly blurred, and mostly by design.", CSM Sheldon A. Watson.

That observation, offered by a defense analyst in 2022, wasn't hyperbole; it was a diagnosis. Most people still imagine warfare as something forged in secrecy: shadowy factories, classified programs, black budgets. But the most transformative weapon of the 21st century, the commercial drone, wasn't born in a bunker. It was born in a marketplace.

Walk into any big-box electronics store and you'll find technology that rivals what the Pentagon once guarded under lock and key. Stabilized cameras. Thermal sensors. Autonomous navigation. All packaged with a smile and a warranty. The irony is staggering: the same components that film family barbecues now guide precision strikes in combat zones. When global supply chains blur borders, commerce becomes the arsenal.

In the 20th century, military hardware came stamped with a flag. "Made in America." "Made in Russia." You knew whose side the tool belonged to. That certainty vanished with globalization. Today's drone is an international collaboration, featuring chips from Taiwan, cameras from Japan, batteries from South Korea, software from California, and assembly in Shenzhen. It's a world economy soldered together on a circuit board. That interdependence makes war harder, and peace riskier. When every side depends on the same suppliers, you can't sanction without self-harm. It's like threatening to cut the power while standing in the same pool. Economically, it's genius. Strategically, it's a nightmare.

In 2014, ISIS released propaganda videos showing commercial quadcopters carrying grenades over the ruins of Mosul. A few years later, the same models appeared in Ukraine and Syria, modified with 3D-printed mounts. None were designed for combat, but war rarely waits for permission. Analysts traced the parts to civilian distributors in Europe and Asia. No espionage. No arms deals. Just online shopping. The digital storefront replaced the black market. A credit card could do what smuggling once did. Conflict went retail. That's the peril of open innovation: every feature designed to empower consumers can be repurposed to put them at risk.

Manufacturers now face an impossible equation: innovate or lose market share, regulate or lose revenue. The global drone industry is projected to exceed $90 billion by 2030. With that kind of profit potential, restraint is rarely the winning strategy. Executives publicly denounce misuse while privately racing to stay ahead of competitors who don't care about misuse. It's a moral math problem with quarterly earnings attached. Some companies have begun embedding digital "geofencing" software that prevents drones from flying in restricted zones, such as airports or government facilities. However, those limits are easily bypassed. Every safeguard inspires an equal and opposite YouTube tutorial. Commerce loves loopholes because conflict thrives in them.

For decades, arms-control treaties kept advanced weaponry from flooding the global market. However, those agreements never anticipated a world where military-grade sensors were packaged in consumer-friendly formats. A camera capable of reading a license plate from 2,000 feet isn't classified; it's advertised. A thermal lens that spots movement through foliage isn't restricted; it's on sale for hunters. So, while defense departments debate procurement cycles, insurgent groups fill online carts. The barrier to entry isn't politics anymore, it's postage.

Copying used to be the sincerest form of flattery. Now it's the fastest form of escalation. Once a drone hits the market, replicas appear within months, cheaper, lighter, and often produced in regions where intellectual-property law is more of a suggestion than a statute. Each clone fragments accountability further. When a modified drone crashes into an oil facility in the Middle East, investigators can't trace the source. Too many middlemen. Too many unbranded parts. The sky becomes anonymous. And anonymity, in the wrong context, is an ally of aggression.

Corporations argue they just sell tools; what people do with them isn't their fault. That defense collapses under scrutiny. History doesn't separate the manufacturer from the outcome; it merges them. Think of chemical companies during World War I, social-media giants amplifying extremism, or banks laundering money for sanctioned regimes. When profit trumps precaution, participation becomes partnership. Today's drone executives stand at that same moral crossroads. They can't claim ignorance when battlefield videos feature their logos. And the public shouldn't let them. Accountability in the 21st century isn't about blame; it's about boundaries. If you can engineer the future, you can also engineer restraint.

Between corporate boards and end users sits a vast ecosystem of distributors, resellers, and online marketplaces. It's the Wild West with better shipping options. Platforms host sellers with minimal verification. Third-party vendors repackage products under new names. Sanctioned actors often use proxies and shell companies to conceal their transactions and activities. The result is a logistical fog where legality dissolves. Even if a manufacturer bans sales to hostile regions, components trickle in through intermediaries because there are no customs checkpoints in cyberspace. The moral of the story: you can't geofence greed.

While corporations accelerate, governments deliberate. By the time policymakers grasp a threat, it's already airborne. Export controls move at diplomatic speed; drone innovation moves at startup speed. Meanwhile, regulators are still updating frameworks from 2016, while models from the 2025 era are learning to think for themselves. This mismatch leaves security agencies patching holes in real time. Law enforcement relies on voluntary corporate cooperation. The military navigates public-relations minefields. Legislators chase headlines instead of root causes. Everyone's busy reacting while no one is truly governing.

Here's the most brutal truth: every innovation multiplies moral responsibility. When a product can cross borders, so does the burden of care. A drone designer in California may never meet the soldier or insurgent who uses their creation, but the connection exists, nonetheless. However, distance doesn't dissolve duty. We tend to romanticize innovation as neutral, "a tool for good or evil." But neutrality is a myth. Every choice of design, distribution, and data policy tilts the scale toward one outcome or the other. The engineer's keyboard is now as consequential as the general's map.

Despite the bleak picture, there's room for optimism. Some companies are embracing traceable supply chains, flight logs encrypted with blockchain, and transparency initiatives that document where and how drones are used. Others partner with defense and humanitarian organizations to standardize safety features and share threat intelligence. Progress occurs when accountability becomes a competitive advantage rather than a compliance chore. The future doesn't require abandoning innovation; it requires illuminating it.

In the next section, *"The Price of Convenience,"* we'll look inward. Because while corporations and nations bear responsibility for production, consumers drive demand. The uncomfortable question is this: what are we willing to trade for comfort? We've built a world where everything arrives faster, cheaper, and smarter, but the bill is coming due.

Section 4: The Price of Convenience

"We shape our buildings; thereafter they shape us," Churchill once said.

If he were alive today, he might revise that line for the digital age: We shape our technology; thereafter, it spoils us. The shift isn't subtle; it's seismic. We've entered the age of effortless everything. Groceries appear on our doorstep, and entertainment streams on demand. Cars park themselves. Drones deliver coffee. Convenience has become the currency of modern life, and we're spending it like there's no inflation.

However, every convenience comes with an invisible cost: dependency. And dependency, when scaled, becomes a liability. We've traded effort for ease, and in doing so, we've surrendered something essential, not just control, but awareness. Every device that simplifies life also collects data. Every automation that saves time stores patterns. Every drone that maps your roof teaches someone, or something, about your home. It's not a conspiracy; it's an equation. We wanted everything to be "smart." We forgot that "smart" means "watching."

The drone that drops your package can also record your property. The camera that tracks deliveries can track you. And the AI that predicts your shopping habits could, in another context, predict your movements. We've built a nation of digital mirrors and stopped asking who's holding the reflection.

Human beings adapt fast, sometimes too fast. We normalize comfort faster than we recognize its consequences. Consider how quickly we transitioned from walking into stores to demanding two-day shipping. Then one day. Now, why not the same afternoon?" We've trained ourselves to expect miracles and panic when they're late. The joke practically writes itself: "We traded situational awareness for same-day shipping." It's funny until you realize it's true. We've outsourced not only labor but vigilance. We've replaced curiosity with convenience. And when technology does everything for you, the first thing it takes is your attention.

History repeats itself, just with faster Wi-Fi. Every civilization that prioritized comfort over caution eventually paid the price. Rome outsourced its defense to mercenaries. The Industrial Age outsourced its conscience to factories. The Digital Age is outsourcing its security to algorithms. We don't notice the shift because comfort feels like progress, until it doesn't. Convenience is never free. Someone, somewhere, pays the cost, often in ways we don't see until it's too late.

Technology gives the illusion of mastery. You push a button, something happens. You order; it arrives. You tap, and the world reacts. But behind that illusion is an ecosystem of dependencies, satellites, networks, code, power grids, and unseen workers, all functioning so seamlessly that we forget they're fragile. When those systems fail, the convenience empire collapses fast. Ask anyone who's lived through a cyberattack, a supply-chain disruption, or a natural disaster: when the lights go out, so does the illusion of control. Drones, for all their wonder, are part of that web. They rely on GPS signals, batteries, and data connectivity. Cut one link, and they fall, literally and figuratively. We built the sky of convenience on a foundation of assumptions.

There's a difference between using technology and leaning on it. The problem arises when the tool becomes the teacher, when we forget how to function without it. Today, modern life is filled with digital training wheels, including navigation apps, fitness trackers, and autocorrect. Drones are the newest addition to the list; they think for us, fly for us, and soon, they may even decide for us. In military circles, we refer to

42

this as automation bias, the tendency to trust machines over human judgment. Civilians have it too. If the app says it's fine, we assume it's fine. But judgment is like a muscle: neglect it, and it atrophies. When people stop questioning, societies stop progressing.

Let's be honest: no one wants to go backward. No one's volunteering to give up convenience. We like our Amazon packages, our automated thermostats, and our aerial footage of backyard barbecues. The goal isn't to reject progress, it's to recognize the trade-offs. Because comfort, unchecked, becomes complacency. And complacency is the one weakness no adversary needs to exploit; we gift-wrap it ourselves. A nation that forgets how to function without automation forgets how to function under pressure.

If aliens ever visited Earth, they'd probably assume drones were the dominant species. Think about it, they fly, they film, they deliver, and humans mostly stand around watching them. That's not science fiction; that's Saturday afternoon in suburbia. We've become spectators in our own innovations. We cheer at convenience like fans at a game, forgetting that someone has to play defense.

Convenience changes not just what we do, but who we are. We're less patient, less curious, and more dependent. We often seek comfort in efficiency rather than excellence. The irony? The same drive that invented convenience was born from discomfort, from people who were hungry, cold, or curious enough to solve problems. We've inherited their tools but misplaced their grit. The challenge now isn't to invent faster, it's to remember why we invent at all.

At scale, the personal cost of convenience becomes a civic one. When a society values ease over effort, its public institutions mirror that attitude. Short-term fixes replace long-term planning. We stop investing in resilience because we assume someone else will handle it. Infrastructure decays. Training budgets shrink. Emergency response gets outsourced to "apps." The same complacency that leaves a neighborhood vulnerable to blackout leaves a nation susceptible to breach. A strong democracy doesn't just need informed voters; it needs alert citizens. And you can't be alert when everything's automated.

None of this means we should fear technology. It means we should mature alongside it. Convenience should serve humanity, not erode it. That starts with awareness, knowing what we're trading when we press "accept." Governments can legislate safety, and corporations can promise responsibility. But only citizens can demand balance. The next phase of innovation isn't about making drones fly longer; it's about making people think deeper. Progress, real progress, isn't convenience perfected; it's consciousness restored.

In the next section, *"Building Guardrails for Genius,"* we'll focus on how to restore that balance: how ethical innovation, adaptive governance, and civic education can keep human hands on the throttle of progress.

Because the future isn't something we watch unfold, it's something we guide. And the sky won't stay friendly by accident.

Section 5: Building Guardrails for Genius

"Technology is a useful servant but a dangerous master," warned Christian Lous Lange.

He wasn't wrong. Every generation stands at a threshold between what it can do and what it should do. Ours just happens to stand with a drone controller in hand. Innovation has never been the problem, its direction. The moral compass that keeps creativity from drifting into chaos. Genius without guidance is a jet without a pilot: impressive, fast, and destined to crash. Building guardrails for genius doesn't mean slowing progress; it means ensuring it has a lane.

We often talk about engineering as a science, but it's also an act of judgment. Every line of code, every circuit design, carries an ethical decision: Who benefits? Who might be harmed? It's time for ethics to sit at the drafting table alongside innovation. This means embedding social impact reviews into every phase of development, not as a compliance checkbox, but as an integral part of the creative process. Imagine if every design meeting ended with the question: "What could go wrong, and how do we prevent it?" That's not pessimism, it's professionalism. Just as architects design buildings to withstand earthquakes, engineers should design technologies to withstand misuse and abuse.

Governments don't need to be faster than innovation; they need to be smarter about governing it. Traditional bureaucracy treats regulation like cement: slow to pour, hard to change. In a world of rapid technological turnover, we need governance that behaves more like clay, flexible, responsive, and reshaped as new realities emerge. Adaptive governance isn't a buzzword; it's a survival skill. That means creating standing review boards for emerging technologies, composed not just of policymakers but of scientists, ethicists, private-sector leaders, and yes, ordinary citizens. Policy should no longer be written in marble; it should be written in pencil, ready for revision as understanding evolves.

For years, the conversation around drones has been framed as a turf war: government versus industry, innovation versus oversight. It's the wrong framing. Security and innovation aren't opposing forces; they're interdependent. A strong regulatory partnership fuels public trust, which in turn fuels adoption. Corporations that embrace transparency and accountability aren't slowing down progress; they're stabilizing it. Consumers are far more likely to trust technology that comes with a moral warranty. When private industry treats ethics as an investment rather than an expense, it not only protects the public but also protects itself.

The most effective guardrails don't come from legislation or litigation; they come from education. A child who learns about digital citizenship today becomes a responsible innovator tomorrow. Drone literacy, data ethics, and civic awareness should be taught with the same urgency as math or science. Schools can run simulation exercises where students debate the policy implications of new tech. Universities can require ethics courses as part of engineering degree programs. Local governments can host "civic tech days" where communities learn about the impact of emerging tools on their daily lives. It's not about scaring the public; it's about preparing them. Awareness creates resilience. Because if the people don't understand the technology, they can't govern it. And if they can't govern it, someone else will.

One of the most dangerous myths of the modern era is that innovation is self-correcting. It's not. Without accountability, every system, political, corporate, or technological, tends to drift toward self-interest. That's why accountability must be baked into every level of progress. Manufacturers should publish transparency reports about product misuse. Governments should streamline channels for the public to report suspicious drone activity. Regulators should collaborate internationally to close loopholes that allow technology laundering through third parties. The world's problems are global, so the guardrails must be too.

For all the talk about autonomy and AI, the future still depends on human judgment. Machines can calculate faster, fly straighter, and process more data, but they can't care. They don't know what it means to protect a community, uphold a value, or draw a moral line. That's our domain. And we can't outsource it. The challenge isn't to make machines more human, it's to make humans more responsible for their machines. That means demanding transparency from manufacturers, caution from lawmakers, and curiosity from citizens. It means remembering that the "Age of Automation" doesn't free us from accountability; it intensifies it.

The temptation with emerging technologies is to wait for someone else to lead, such as a treaty, a task force, or an international consensus. But leadership doesn't trickle down; it radiates outward. America can model what ethical innovation looks like by holding itself to a higher standard:

- Open-source transparency on public drone data.
- Incentives for ethical research and manufacturing.
- International partnerships on counter-drone and privacy protection.

If we lead with integrity rather than fear, others will follow. If we wait, others will define the rules for us. Guardrails don't limit greatness; they preserve it.

Progress has always been a balance between ambition and humility. Lately, ambition's been winning by a landslide. Humility doesn't mean fear of failure; it means awareness of consequences. It's the difference between thinking you're in control and knowing you're a custodian. Every innovator should remember that the tools they create will outlive them. Their legacy won't just be measured in patents, but in the problems those patents solve or cause. A drone that saves a life and a drone that ends one both trace back to human hands. However, which legacy we build depends on whether we teach humility alongside ingenuity.

Every technological age forces us to confront an ancient truth: wisdom must keep pace with power. Scripture warns of this dynamic often, from the Tower of Babel to the Book of Proverbs. Human pride builds high, but without wisdom, the structure collapses under its own ambition. Faith, regardless of denomination, offers a timeless counterweight to technological arrogance: the idea that creation requires stewardship. We're not gods with gadgets, we're guardians with responsibilities. When innovation aligns with virtue, progress becomes preservation.

We've reached a point where technology can do almost anything, except decide why it should. That decision belongs to us. Guardrails for genius are not barriers; they're boundaries that keep civilization from turning brilliance into collateral damage. The drone age doesn't have to mirror the nuclear age, but without deliberate restraint, it will. We've already proven we can innovate at the speed of imagination. The next test is whether we can mature at the same pace.

In the next chapter, "*The New Battlespace: Homeland Defense Reimagined*," we'll shift focus from innovation itself to the changing nature of defense in a drone-saturated world. We'll explore how the definition of "battlefield" now includes your backyard, your power grid, and your phone signal, and how homeland security must evolve from a fortress mentality to an adaptive network. Because protecting the nation isn't just about building stronger walls, it's about creating smarter skies.

Chapter 3 – The New Battlespace: Homeland Defense Reimagined

Section 1: The Changing Face of War

The morning was clear, the kind of blue Colorado sky that makes you forget anything dangerous could exist above it.

At 8:12 a.m., a technician at a suburban energy substation noticed something hovering in the distance. A drone. Nothing fancy. Just a silver quadcopter, silent and slow. He assumed it belonged to the inspection crew. It didn't.

The facility's flight schedule showed no authorized flights. The drone lingered for seven minutes, circling methodically, tilting its camera toward transformers, relay boxes, and entry gates. Then it zipped away, southbound, at an incredible speed. By the time law enforcement arrived, there was nothing to see, no pilot, no launch site, no trace except for a blurry reflection on one security camera. The report would later list the incident as "unresolved." But that moment, calm, quiet, over in less than ten minutes, marked something profound: the modern face of warfare. Not shock and awe. Not tanks or troops. Just curiosity with a flight path.

"The future of war is not about territory, it's about access." That insight, offered by a NORTHCOM officer, captures the shift with chilling precision. For most of American history, war had a clear set of coordinates. It happened somewhere else, Normandy, Fallujah, or Kandahar. You knew when you were in the fight and when you weren't. Those days are gone. Today, the battle can arrive by signal, sensor, or drone, sometimes without anyone declaring it. The world's most powerful militaries can be disrupted not by fleets, but by firmware. The battlespace has expanded to include every domain technology touches: air, cyber, space, and, most recently, the 400 feet above our neighborhoods. This isn't science fiction, it's strategy in transition.

The drone embodies a new style of conflict, one in which confrontation no longer requires physical contact. In Ukraine, $500 drones have destroyed multimillion-dollar tanks. In the Middle East, non-state actors use off-the-shelf quadcopters to harass troops or monitor supply routes. In the United States, drones have flown over power plants, airports, and correctional facilities. None of these is random. They're all connected by the same principle: leverage through low cost and low risk. Once, only nations could project power from a distance. Now, anyone with Wi-Fi and intent can do it. That changes everything, especially how we define "defense."

During the Cold War, peace relied on deterrence, the belief that some weapons were so catastrophic that no one would dare use them. The drone era shattered that illusion. How do you deter an enemy whose tools are cheap, replaceable, and often untraceable? You can't retaliate when you can't identify who's flying the device. Deterrence depends on fear of consequences. However, the new battlespace operates on the comfort of anonymity. That's why the threat isn't just tactical, it's psychological. When the enemy can strike without warning, attribution, or even presence, defense becomes less about response and more about resilience.

It's tempting to think war remains confined to soldiers and specialists. But technology doesn't recognize those boundaries. The modern soldier and the civilian now share the same sky. The same drone that supports disaster relief can also be used for surveillance in a police precinct. In 2020 alone, U.S. law enforcement agencies reported over 2,000 unauthorized drone incursions near sensitive facilities, a 300% increase from the previous five years. Most were dismissed as hobbyist errors. However, pattern analysis revealed that some flights followed the same paths at suspiciously regular intervals. That's not curiosity. That's reconnaissance.

The disturbing part isn't the number, it's the normalization. People see drones overhead and assume they're harmless. But if every camera on the ground could suddenly fly, privacy, security, and sovereignty would all be up for grabs. We're not just watching the battlefield evolve; we're standing in it.

Distance used to equal safety. That's what made America feel untouchable: two oceans, vast borders, layers of defense. However, in the drone era, distance is no longer a significant factor. A small aircraft can launch from a boat, a border town, or a backyard and reach critical infrastructure in minutes. It doesn't need to cross a checkpoint; it just needs to clear the tree line. The digital layer compounds this. With remote piloting, the "operator" could be ten miles away, or ten thousand. You can't bomb coordinates when the enemy lives in the cloud. That's the new paradox: we're the most connected society in history, and therefore, the most exposed.

In warfare, cost and consequence are often correlated. The more destructive the weapon, the more expensive it was. That rule no longer applies. A commercial drone capable of carrying explosives costs less than a family dinner. And when it's lost, there's no strategic setback, just a restock order. That asymmetry gives smaller actors enormous leverage. Why build a billion-dollar fighter jet when a swarm of $500 drones can achieve the same effect at a fraction of the cost? Power has been democratized, and with it, chaos has ensued.

Every drone, even an innocent one, is a flying sensor suite, comprising a camera, GPS, data link, and processor. Multiply that by millions of units, and the result is a sky that sees more than it ever has before. The question isn't whether that data exists; it's who owns it. When private drones capture footage of critical infrastructure or sensitive locations, that information often ends up on cloud servers with little oversight. Intelligence agencies abroad no longer need to infiltrate networks; they can just collect the byproducts of our curiosity. We've built the most comprehensive reconnaissance network on Earth and made it crowd-sourced.

Notably, America's strategic posture still operates as if conflict originates "over there." Our training, our funding, even our mindset assumes the homeland is the prize, not the playing field. That lag is dangerous. Every adversary who studies us is aware of it. They understand that our greatest vulnerability isn't technology, it's assumption. We assume geography protects us. We assume treaties deter them. We presume our systems will recognize danger in time. However, the sad truth is that assumption has never been a defense strategy.

If every inch of the sky is contested, then every citizen becomes part of the defense. That doesn't mean militarizing the public; it means mobilizing awareness. Homeland defense now includes energy workers, first responders, and even recreational pilots. The "soldier" isn't always in uniform anymore; sometimes, it's the person who decides to report a suspicious flight instead of ignoring it. The future of security isn't about control; it's about collaboration. When ordinary people understand that vigilance is a shared duty, deterrence takes on a new form. Not one of fear, but function.

This shift requires more than new technology; it demands new thinking. Leaders must redefine defense not as a fortress, but as a network of resilience, one where communication, coordination, and civilian readiness matter as much as weapons and walls. That mindset starts at the top but succeeds at the bottom. Because the next battle may not begin with an invasion or explosion, it may begin with silence, signal loss, or a flicker on a monitor. Preparation doesn't mean paranoia; it means refusing to be surprised.

In the next section, *"Civilian Airspace: The Next Frontline,"* we'll look closer at where this new war is already unfolding: the skies directly above our cities, neighborhoods, and critical sites. In the 21st century, airspace isn't just regulated; it's also weaponized. And if we don't learn to manage it as a shared battlefield, someone else will.

Section 2: Civilian Airspace: The Next Frontline

"Airspace maps, rather than traditional battle plans, define the frontlines of future conflicts.", CSM Sheldon A. Watson.

That insight from an FAA safety officer isn't just poetic, it's prophetic. For most people, airspace feels abstract, an invisible ceiling policed by radar and acronyms. The truth is more chaotic. Above 60,000 feet, the domain belongs to satellites and supersonic jets. Between 18,000 and

60,000 feet, commercial aviation rules apply. But below 400 feet? That's where the new frontier lives, unstructured, under-supervised, and increasingly exploited.

The FAA regulates flight safety, but not intent. DHS monitors threats, but not hobbyists. Local police can respond to trespassing, but not always to drones. Somewhere in that 400-foot band of confusion, the nation's most porous battlefield emerged.

Every pilot knows the rule: drones can't fly higher than 400 feet without clearance. The problem is that enforcement is nearly impossible. Traditional radar systems were designed to track aircraft with transponders, not drones smaller than seagulls. Low-altitude radar coverage is patchy, especially around cities, where signal clutter can hide small objects. The result? A vast gray zone, legal in theory, invisible in practice. That's why, when a drone appears near an airport or power plant, it's often spotted by the naked eye, not by sensors. Detection is human; deterrence is optional. We built a regulatory ladder to the clouds and forgot to guard the first rung.

Ask ten agencies who owns the sky under 400 feet, and you'll get eleven answers. The FAA says it governs the airspace. Local law enforcement states that it is responsible for public safety. The Department of Defense insists it can't act domestically without specific authorization. Meanwhile, threats don't wait for legal memos. This "jurisdictional whiplash" leaves responders in a bind. A police officer can see a drone near a stadium, but may not have the authority or tools to take it down. A base commander might detect an incursion but lack jurisdiction to act unless it poses a direct military threat. In other words, the bad guys fly in loopholes, not airspace.

The FAA's mission is safety, not security. It ensures that things in the air don't collide, not that they don't spy, smuggle, or sabotage. That's like running traffic control for a highway without enforcing speed limits. When drones entered the civilian market, the FAA tried to adapt. It introduced Remote ID, a digital license plate for unmanned aircraft. But implementation remains slow, enforcement is inconsistent, and compliance is voluntary for many models. Also, technology evolves faster than paperwork, and the paperwork was written for airplanes. So, while the FAA debates regulatory nuance, the airspace beneath it fills with noise, and not all of it is friendly.

The commercial drone market is growing at a rate of nearly 15% per year, adding thousands of new aircraft to already congested skies. Currently, most operators follow the rules. Some don't. A few never intended to. And so, the difference between a delivery drone and a reconnaissance drone often lies in the software, which is easily rewritten. Yet the agencies responsible for oversight operate on 20th-century timetables, with budget cycles, legislative reviews, and inter-agency task forces. By the time the rules are agreed upon, the technology has already outgrown them. Oversight without agility is theater.

Ask an average citizen who's responsible for protecting their local airspace, and you'll likely get a shrug. Most assume the government has it handled. Some think the military can act anywhere. Others believe "someone up there" is watching. That blind faith in authority might've worked in the 1950s. It doesn't hold in the drone age. Today, the airspace over your neighborhood isn't just shared, it's contested. A curious teenager, a contractor, or a hostile actor could all use the same device, in the same sky, with the same impunity. Without public literacy, every drone looks innocent until it isn't.

Hospitals, refineries, power substations, and data centers were never designed for aerial defense. Most have fences, cameras, and guards, but nothing that looks up. That oversight made sense when threats arrived on foot or wheels. It doesn't when they descend silently from the clouds. A drone carrying a small payload, even a brick, can disrupt critical systems, not because of its destructive power, but because of its unpredictability. One well-timed incursion can trigger lockdowns, evacuations, or cascading power failures. The softest targets aren't undefended; they're unprepared.

While national security experts debate sabotage, ordinary Americans are grappling with something more personal: privacy. Homeowners spot drones hovering over backyards. Parents worry about schoolyards being filmed. Police departments grapple with balancing surveillance against civil liberties. Food for thought, the same camera that captures wedding footage can capture license plates, faces, or private property. If the public begins to see every drone as a potential threat, the legitimate industry will suffer. Thus, the line between curiosity and criminality has never been thinner or more confusing. That's why public education isn't just good policy, it's preemptive diplomacy.

There's a saying among drone operators: "If you're flying legally, no one notices. If you're flying illegally, everyone suddenly becomes an expert." It's funny because it's true. Americans ignore the sky until something goes wrong. Then everyone looks up like they've been on the lookout for years. But humor aside, that reaction gap, apathy before, panic after, is precisely what adversaries exploit. The more predictable our response, the easier it is for us to be manipulated. Awareness is the most cost-effective and renewable countermeasure we have.

Bridging the airspace gap requires three layers of alignment: policy, technology, and people.

Policy: Clarify who's in charge below 400 feet, and adopt a unified framework that empowers local responders without paralyzing them with red tape.

Technology: Deploy low-altitude detection systems that integrate civilian and defense networks in real time.

People: Educate the public to recognize, report, and operate drones responsibly.

The solution isn't more regulation, it's smarter coordination. The sky isn't anyone's sole jurisdiction; it's everyone's shared responsibility.

Air superiority once meant controlling the clouds. Now it means managing the space above the treetops. Victory in this new battlespace won't come from shooting down threats; it'll come from seeing them first, understanding them faster, and responding without hesitation. That requires partnerships among federal, state, local, and civilian entities. The front lines are everywhere, which means defenders must be, too.

The future of homeland defense won't be written in air-defense manuals alone; it'll be written in city ordinances, FAA memos, and neighborhood watch briefings.

In the next section, *"The National Airspace System's Soft Underbelly,"* we'll examine the infrastructure holding all this together: the radar grids, communications networks, and energy systems that make the modern sky possible, and how their weaknesses have become open invitations to exploitation. Because before we can secure the sky, we have to admit how fragile it truly is.

Section 3: The National Airspace System's Soft Underbelly

"Amateurs talk strategy. Professionals talk logistics.", *Gen. Omar Bradley.*

Gen. Omar Bradley's words echo louder than ever in the age of drones. Most people think airspace security is about interceptors, radars, and no-fly zones. But the real battlefield lies in the machinery that enables flight. The National Airspace System (NAS) is one of the most complex man-made networks on Earth, comprising 5,000 public-use airports, 13,000 radar feeds, 20,000 air traffic controllers, and over 45,000 flights per day. It's an orchestra of precision. But like any orchestra, it's only as good as its weakest instrument. And right now, that instrument is low-altitude awareness, the blind spot drones were born to exploit.

From the ground up, America's airspace is layered like a wedding cake, each tier designed for a specific altitude, aircraft type, and level of control.

Class A: Above 18,000 feet, the realm of commercial airliners.

Class B/C: Surrounds major airports.

Class D: Regional fields.

Class E/G: Everything else.

This system worked flawlessly when every flyer filed a flight plan, and every plane had a transponder. Drones broke that logic. They're small, cheap, and often unregistered. They fly below radar coverage and outside predictable flight paths. They are, by design, ghosts in the system. That means the NAS, the digital nervous system of American airspace, has an exposed nerve no one designed it to feel.

Traditional radar is a marvel of physics and patience, but it was built for a world of large, metallic objects at high altitude. A modern drone? Too small. Too low. Too quiet. Radar waves bounce off trees, buildings, and weather at those heights. The result: clutter so thick that small drones become invisible within it. Some advanced systems use acoustic or optical sensors to supplement radar, but coverage is inconsistent. Urban areas are complicated. Skyscrapers create reflections. Concrete blocks the line of sight. Interference buries signatures. The same concrete jungle that protects us from view hides our vulnerabilities from detection.

After 9/11, America poured billions into overlapping defense layers, NORAD, FAA, DHS, FBI fusion centers, and countless classified networks. From the stratosphere down to the ground, sensors scan, listen, and analyze. However, those systems were designed to identify large, fast-moving threats, not backyard technology that costs less than a laptop. We have the most advanced surveillance grid in the world, yet a $300 drone can slip past it unnoticed. That's not a failure of intelligence; it's a failure of imagination. As threats get smaller, our response systems need to get smarter. Size doesn't equal significance anymore.

In theory, the United States has a comprehensive view of its skies. In practice, that picture resembles a patchwork quilt, stitched together by agencies that rarely share data in real-time. Each sector, military, civil, and commercial, operates its own sensors, databases, and alert systems. But integration is still limited by bureaucracy, classification levels, and technical incompatibilities. In other words, the data exists, but it just doesn't talk to itself. That's like having the pieces of a radar jigsaw puzzle but never putting them on the same table.

NAS doesn't just rely on machines; it depends on networks. And networks are inherently vulnerable. Cyberattacks on GPS signals, radar feeds, or communication relays could paralyze air-traffic management within minutes. It has already occurred on small scales: GPS spoofing incidents in the Middle East, jamming attacks near airports, and the transmission of false altitude data into aviation systems. If you can fool a sensor, you can control a response. The more connected the system becomes, the more exposed it is. Connectivity is power, and also a single point of failure.

For decades, NORAD has guarded North America's skies with vigilance and precision. But NORAD's radar grid doesn't extend far below 3,000 feet. Its mission was always strategic, to intercept bombers, not hobby drones. Below that altitude, responsibility is divided among agencies. The FAA ensures flight safety. DHS manages homeland security threats. Local law enforcement responds to incidents as they arise. In reality, no one owns the low sky, so no one can fully defend it. The enemy no longer needs stealth aircraft. They need smallness and timing.

Imagine the United States mapped from above: thousands of overlapping circles representing restricted zones, airports, military bases, nuclear plants, VIP residences. Now imagine everything in between: vast, unmonitored corridors where drones can move freely. Many critical facilities are situated just outside those no-fly zones. A drone flying a few hundred feet beyond the perimeter is still technically legal and tactically practical. The security model built on "perimeter protection" doesn't work when the sky has no fences. A power station can install razor wire, but it can't install radar.

Most people think system failure means catastrophe, alarms blaring, lights flashing, chaos. However, in airspace defense, failure appears as silence. When a drone flies undetected, there's no alarm to sound. When it captures footage of a substation, there's no alert to trigger. When it drops contraband into a prison yard, it's over before anyone looks up. That's the danger of invisible threats; they don't announce themselves. And because most of these incidents cause no immediate damage, they rarely make headlines, which means they seldom drive reform.

The irony is brutal to ignore: the same American innovation that built the air traffic control system is now outpacing it. The FAA, NORAD, and DHS still work with infrastructure designed before smartphones existed. Meanwhile, private companies test drones capable of autonomous navigation, swarming, and AI-powered decision-making. Defense systems evolve through appropriations and contracts. Threats evolve through firmware updates. Until those timelines converge, we'll keep losing ground quietly.

Engineers joke that the air traffic control system is like an old battleship: sturdy, dependable, and always in need of a patch. The joke works because it's true, and because no one wants to admit how old some of those "patches" are. It's like running modern software on a rotary phone. However, here's the punchline: it still works primarily. And that's both the comfort and the curse because success breeds complacency, until the day it doesn't.

Fixing the soft underbelly doesn't mean replacing the system; it means reinforcing it. That starts with integration: linking civil, military, and private-sector sensors into a shared network. Then, automation: using AI to filter clutter and detect anomalies in real-time. Ultimately, adaptation entails developing legal and tactical frameworks that facilitate swift responses without bureaucratic delays. A smarter sky isn't just about detection; it's about coordination. The threat doesn't respect jurisdiction; neither should the defense.

In the next section, *"Drones and Critical Infrastructure: The Achilles' Heel of Modern Life,"* we'll zoom in further to explore how this fragile airspace system intersects with the physical lifelines that keep the nation running: power, water, transportation, and communication. Because protecting the sky means protecting everything beneath it, and right now, both are far more vulnerable than we'd like to admit.

Section 4: Drones and Critical Infrastructure: The Achilles' Heel of Modern Life

"If you want to cripple a nation, you don't need to bomb its cities, just unplug them.", CSM Shelon A. Watson.

That warning from a cyber defense analyst captures the essence of modern vulnerability. The most powerful nations in history weren't toppled by invasion; they were undone by interruption. Rome lost its roads. The Soviets lost their economy. And so, the modern world could lose its infrastructure and never even realize it.

We live in a civilization of dependencies: power grids, pipelines, data centers, and distribution networks, all interconnected, all automated, all vulnerable from above. A single well-placed drone doesn't have to destroy anything. It just has to disrupt. Disruption is the modern equivalent of destruction.

Start with the grid, the circulatory system of modern life. In 2022, an unauthorized drone flew near a Pennsylvania power plant, capturing detailed footage of transformers and transmission lines. The pilot was never identified. No damage occurred, but cybersecurity analysts flagged the incident as a "pattern reconnaissance event." That's bureaucratic language for "someone was mapping your weak spots." The energy sector faces an average of ten drone incursions per month nationwide; most are logged, but few are investigated. Many facilities rely on passive deterrence: signs, fences, and the hope that they will be effective. A quadcopter with a $30 magnet can short a transformer. A drone carrying a thin wire can cause cascading outages. It's not science fiction, it's physics. Our grid is built for weather, not warfare. And the weather now has propellers.

Water-treatment plants are among the most neglected security sites in America. Most operate on thin budgets, aging infrastructure, and limited staff. In 2021, an unidentified drone hovered over a Florida water facility for several minutes, recording intake valves and sensor equipment. A week later, the same plant was targeted by a cyberattack aimed at altering chemical levels. Coincidence? Possibly. But probability doesn't comfort the thirsty. If energy is the pulse, water is the bloodstream. A contamination event, even a localized one, could paralyze a city and shatter public trust in an instant. Unlike oil or gas, there's no stockpile for water. When it stops, everything else stops as well.

America's highways, bridges, railways, and airports are engineering marvels and surveillance magnets. Drones have been spotted over major interstates, recording traffic flow and response patterns after accidents. Others have loitered near airport approach paths, prompting flight delays. In 2023, a drone nearly collided with a passenger jet near Boston Logan, one second of hesitation away from tragedy. Then there's rail. Freight trains transport a wide range of goods, from chlorine to crude oil. A drone doesn't need explosives to cause harm; it can simply provide coordinates. Intelligence, after all, is the most dangerous payload. Our infrastructure moves faster than our ability to monitor it. Every bridge, runway, and terminal is both an asset and a target.

The Internet isn't in the clouds; it's in cables, satellites, and towers. Those towers are tall, exposed, and often unguarded. Telecommunication hubs connect entire regions, yet most are protected by nothing more than chain-link fences. A drone carrying conductive wire or flammable material could disrupt service for thousands, possibly millions. And because data defines modern life, from banking and logistics to healthcare and communications, a communications failure isn't just an inconvenience; it's paralysis. In 2020, a drone carrying a copper wire was discovered near a cell tower in Nashville. Officials dismissed it as "possible vandalism." They missed the point. In an interconnected world, even vandalism can be a rehearsal.

Unfortunately, few realize how dependent the food chain has become on aerial technology. Farmers use drones for irrigation mapping, crop analysis, and pest control. That's progress, until someone hijacks the same model to disrupt or contaminate. An agricultural drone spraying fertilizer looks identical to one dispersing chemicals. A drone surveying farmland can also serve as a surveillance tool along supply-chain routes. Agriculture is now part of homeland defense, though it doesn't know it yet.

None of these threats requires state actors. They require imagination and access, two things the modern marketplace provides in abundance. The "dual-use dilemma" hits hardest here: tools meant for industry and innovation can easily become instruments of exploitation. A drone designed to deliver medicine can also deliver a payload. A drone used for photography can also be used for espionage. A drone used for crop dusting can also disperse toxins. It's not about capability, it's about intent. And intent is impossible to regulate.

Why is this problem so persistent? Security rarely competes well against convenience or cost. Securing a single power plant against drone threats can cost millions, while a drone attack might cost hundreds. That imbalance discourages proactive defense. Private companies prioritize efficiency. Public agencies prioritize jurisdiction. Adversaries prioritize results. Yes, it's an unfair fight, but it's one we can win with coordination, not cash.

Protecting infrastructure in the drone age requires a new doctrine: layered defense.

- **Awareness**: Real-time detection through radar, acoustic, and visual sensors.

- **Access Control**: Designated no-fly zones dynamically updated with local threats.

- **Active Deterrence**: Non-kinetic countermeasures like signal jamming or drone capture systems.

- **Response**: Trained personnel with clear legal authority to act.

- **Resilience**: Systems that can absorb and recover from disruption quickly.

This isn't about militarizing the homeland; it's about modernizing it. When each layer works, communications and attacks lose impact before they gain momentum.

In military terms, we call this "defending the donut", everything around the hole while ignoring what's inside it. America has spent decades protecting the perimeter, borders, bases, and skies, while assuming the interior was safe by default. Now, the hole is the target. And the hole isn't empty, it's full of substations, hospitals, servers, and pipelines. You can't defend what you don't see. And until recently, we didn't even look.

Defense of critical infrastructure demands collaboration, not command. The DOD can't guard every substation. The FAA can't monitor every drone. The DHS can't respond to every incident. But together, with industry partners, local responders, and informed citizens, they can build a net strong enough to catch threats before they fall through. That's the new definition of national defense: distributed vigilance. Every responder, every technician, every neighbor is part of the detection network. The more eyes that look up, the safer the ground becomes.

In the final section, *"The Blurred Line: Law Enforcement vs. Military Jurisdiction,"* we'll confront the most challenging part of this new battlespace: who's actually in charge when the sky turns hostile. Because in a world where every drone could be a weapon, the line between public safety and national defense isn't just blurred, it's fading fast.

Section 5: The Blurred Line: Law Enforcement vs. Military Jurisdiction

"The Constitution is not a suicide pact," Justice Robert H. Jackson warned in 1949.

He was speaking about the tension between liberty and survival, a tension that now hovers, quite literally, above our heads.

When a drone enters restricted airspace, who's in charge, the cop on the ground or the commander in the sky? The question sounds simple until you try to answer it. If the drone poses a threat to life or national security, logic says the military should intervene. But if it's operating domestically, the *Posse Comitatus Act* limits direct military involvement in civilian law enforcement. The result is a jurisdictional standoff: a threat that demands immediate action in a space where authority moves at legal speed. The adversary's advantage isn't firepower; it's ambiguity.

Imagine this: a drone breaches restricted airspace above a downtown event. Local police can see it, but they cannot identify it. The FAA is notified, but has no real-time response capability. DHS tracks it on sensors but lacks immediate enforcement power. NORAD is alerted, but its jurisdiction technically begins above the city's rooftops, not between them. By the time everyone confers, the drone has vanished. That isn't incompetence, it's fragmentation. The system wasn't built for speed; it was built for hierarchy. And so, adversaries thrive in the space between silos.

The Posse Comitatus Act of 1878 was enacted to protect civil liberties by keeping the military out of domestic law enforcement. That principle is sacred, but it's also showing its age. In 1878, threats came on horseback, not over the internet. The framers of that law couldn't have imagined AI-guided drones or state-sponsored cyber intrusions that blend into civilian life. We now face enemies that operate inside our legal frameworks, exploiting our own restraint as a strategy. The challenge isn't whether to rewrite the law, it's how to interpret it responsibly in a world it was never designed for. Freedom and defense aren't mutually exclusive. But they do require coordination.

The Department of Homeland Security and the Department of Defense maintain a delicate partnership, each with overlapping missions and non-overlapping authorities.

DHS protects the homeland from domestic threats.

DOD protects the nation from external threats.

But drones don't respect those boundaries. A device launched from an American backyard can be controlled from another continent. That leaves defenders caught in a bureaucratic Bermuda Triangle, where every actor has partial authority, but no one has complete control. Until those boundaries are redrawn, every response will start with confusion.

Unquestionably, local police are often the first to see a threat, but the last to get clearance to act. Most departments lack counter-UAS equipment and training. Even if they had the tools, federal law restricts them from jamming or disabling drones, actions considered interference with federal airspace. In practice, officers can observe, report, and wait. That delay is deadly in a domain where threats move faster than the speed of paperwork. This isn't a criticism of law enforcement; it's an indictment of the system that ties their hands with red tape and good intentions.

The military, meanwhile, has the expertise, sensors, and kinetic capability to respond. But domestically, it must tread lightly, politically and legally. No one wants armed soldiers stationed at airports or city halls. The American people trust their military, but fear its presence in civic life. That tension is healthy until it becomes hazardous. We need a framework for domestic defense that empowers response without eroding public trust. The enemy shouldn't be the reason we abandon our own values.

In recent years, Congress has granted limited authority to certain federal agencies, such as DHS, DOJ, and DOD, to engage in counter-UAS activities under specific conditions. But those authorizations are narrow, temporary, and complex. For example:

The military can engage drones on base, but not necessarily over it if they cross into civilian airspace.

Law enforcement can report a threat, but can't neutralize it without federal approval.

The FAA can declare restricted airspace, but can't enforce it kinetically.

That's like having fire departments that can't use water unless Washington signs off. Legislation hasn't kept pace with reality. The sky is changing faster than the statutes that govern it.

Even if authority were apparent, execution requires restraint. Shooting down a drone isn't like stopping a car; it falls, often unpredictably. One wrong intercept could cause collateral damage. Then there's the privacy dimension: distinguishing a malicious drone from a journalist's, or a hobbyist's, isn't always immediate. Overreaction risks public trust; underreaction risks public safety. In short, we're defending democracy while trying not to dent it. That balancing act isn't easy, but it's necessary. The moment national defense becomes indistinguishable from domestic control, we've lost the very thing we're trying to protect.

The answer isn't to militarize police or civilianize soldiers, it's to train together. Joint counter-UAS exercises should be as routine as active-shooter drills or cyber-resilience tests. Shared terminology, shared tools, shared expectations, that's how you close the gap between awareness and action. A drone doesn't care what patch is on your uniform. Neither should the response. This kind of interoperability isn't just tactical, it's cultural. It reminds every participant that the mission isn't ownership, it's defense.

Ultimately, homeland defense depends on consent. The public must believe that those guarding the skies are accountable, disciplined, and lawful, because transparency matters. When people understand why specific actions are taken, or not taken, suspicion fades and cooperation grows. Trust forms the foundation of every effective domestic security program, including neighborhood watch initiatives and cyber awareness efforts, and increases through collective engagement. Without it, even the most advanced counter-UAS system will fail, because security isn't a system; it's a relationship.

There's an old saying in the Army: "If everyone's in charge, no one's in charge." That might as well be the motto for domestic drone response. Every agency has authority, right up until the moment it matters. Then everyone politely looks at each other and says, "That's not our lane." Meanwhile, the drone doesn't wait for consensus.

The solution isn't another law or acronym, it's a philosophy: Unity Without Overreach.

Unity means interoperability, shared data, protocols, and training.

Overreach means remembering that civil liberties are not collateral.

We need a national counter-UAS framework that does three things:

- Clarifies authority, who can act, when, and under what conditions.
- Empowers local responders with tools and training, not just responsibility.
- Reassures the public that safety and liberty can coexist.

That framework doesn't weaken democracy. It reinforces it.

As someone who's worn the uniform long enough to watch technology rewrite tactics, I'll say this plainly: the fight has already arrived home. We just haven't adjusted our posture. The future of homeland defense won't look like checkpoints or patrols; it'll look like partnerships, data-sharing, and disciplined awareness. Our job isn't to prepare for the war we fear. It's to prepare for the one we're already in, the silent kind that hovers, records, and waits.

In the next chapter, *"Invisible Wars: The Domestic Frontline,"* we move from what flies above us to what unfolds among us. The battlefield no longer begins at foreign borders; it starts at home, in neighborhoods, city grids, and unseen networks connecting every device and every person. The enemies of the modern age no longer rely on fleets or flags; they exploit confusion, complacency, and the spaces where vigilance has fallen silent. Because while drones occupy the air, the real battle is for what people believe, for the trust that holds a nation together and the awareness that keeps it alert. And in that fight, the most potent weapon isn't firepower, it's understanding.

Chapter 4 – Invisible Wars: The Domestic Frontline

Section 1: The Comfort of Not Knowing

"Wars used to be declared; now they just arrive.", **CSM** *Shelon A. Watson.*

That insight from a DHS analyst in 2023 wasn't a metaphor; it was a memo. The battlefield has changed, not because we invited war, but because we connected everything.

It started with a call to dispatch. A drone delivery system in suburban Phoenix misread a newly updated no-fly zone after a construction permit changed. The drone rerouted, not to safety, but over an elementary school during recess. Children pointed skyward as the box-laden quadcopter hovered silently, unable to decide whether to land, return, or wait. Local law enforcement scrambled to contact the vendor. The vendor scrambled to contact the algorithm vendor. And the algorithm? It waited for better GPS. Nobody was harmed. But the moment revealed something more profound: the system moved not with malice but without accountability. It wasn't evil. It was just autonomous.

It doesn't take an explosion to start a war anymore. All it takes is an intrusion, a drone flying low over a substation, a sensor misreading, a camera quietly capturing what should never be seen. Most Americans will never hear the buzz, never notice the flash of plastic and carbon fiber above their rooftops. Yet each flight could represent a rehearsal, a probe, or a dry run for something larger. The homeland, once a sanctuary behind the oceans, has become a silent battlefield. Invisible warfare has no uniforms, no frontlines, and no clear enemy. It slips between the seams of daily life, disguised as innovation or convenience. That's why the greatest vulnerability of the twenty-first century isn't technology, it's the illusion of safety.

For decades, America's defense mindset was outward, focusing on oceans, borders, and perimeters. But now, the threat doesn't come through the gate; it floats above it. Over seventy percent of the United States is classified as low-altitude, unrestricted airspace. That means nearly every suburban neighborhood, industrial park, and power corridor lies within reach of a commercially available drone. In theory, the FAA governs that airspace. In practice, it can't monitor more than a fraction of it. Enforcement relies on radar overlays, visual observation, and public reporting, an approach designed for pilots, not swarms. We once feared enemies crossing borders. Now, the most dangerous intrusions cross no line at all.

Drones have become the modern Swiss Army knife, adaptable, accessible, and misunderstood. What began as a tool for hobbyists and cinematographers quickly evolved into an ecosystem of surveillance, smuggling, and sabotage. The problem isn't the drone, it's the operator. At first, incidents seemed harmless: tourists filming landmarks, real estate agents capturing aerial shots, farmers surveying fields. But beneath those legitimate flights were others with darker intent. A drone hovering over a military base might belong to a journalist, or it might be mapping heat signatures for a foreign buyer. Again, the same model that films weddings can also deliver payloads. The same software that tracks crops can track convoys. The American sky, once defined by commercial aviation, now resembles an unregulated marketplace of motives.

In December 2019 and January 2020, residents across eastern Colorado and western Nebraska began reporting fleets of drones flying in formation. Witnesses described synchronized movement, steady altitude, and silent precision. The flights occurred nightly for weeks. Law enforcement coordinated patrols, radar checks, and FAA alerts. No organization claimed responsibility. No pilot was ever identified. To this day, the "Colorado drone swarm" remains unexplained. Some insist it was a private experiment. Others suspect foreign reconnaissance. Whatever it was, it exposed a simple truth: America was unprepared to answer the question, "Who owns the sky tonight?" It also revealed something more profound: fear thrives in ambiguity. Without clear communication or response, rumors became their own weapon. That was the first modern domestic drone panic, and it will not be the last.

When a drone is spotted near critical infrastructure, who responds first? The FAA? DHS? Local law enforcement? The answer, too often, is "no one fast enough." Jurisdiction overlaps, but responsibility diffuses. FAA governs airspace. DHS handles homeland threats. Local police handle public safety. DoD manages defense installations. Each agency has a mandate, but none has authority to act unilaterally in mixed-use civilian airspace. This bureaucratic fragmentation turns minutes into hours. And in drone warfare, minutes are eternity. Until Congress and the executive branch reconcile authority gaps, the first response will remain confusion, and the first casualty will be time.

In 2022, a drone dropped contraband into a prison yard in Georgia. In 2023, a collision involving a police helicopter occurred in Los Angeles. In 2024, another nearly struck a passenger aircraft as it approached Austin-Bergstrom International Airport. Each event ended without catastrophe. But each exposed the same pattern: proximity without consequence. Every warning becomes a case study, then fades from public attention. However, these minor breaches accumulate into a doctrine: test the limits until someone notices. Soft targets, such as schools, churches, and sports venues, remain vulnerable. Unlike cyber threats, there is no firewall for physical airspace. The only defense is vigilance. The tragedy is that vigilance has become increasingly rare in a culture that is increasingly addicted to convenience.

Domestic security professionals often describe the homeland as a "soft network", one with strong institutions but weak coordination. In the drone era, that imbalance becomes a liability. Imagine a drone swarm simultaneously targeting substations across three states. The physical damage might be minimal. But the psychological effect would be devastating. Power outages, news reports, and social media speculation, all feeding confusion faster than the system can clarify. Adversaries don't have to destroy infrastructure. They only have to erode trust in its reliability. That is how invisible wars work: they turn uncertainty into strategy.

Ask ten Americans who governs their airspace, and most won't know. Ask how to report a suspicious drone, and they'll guess. Public ignorance isn't just inconvenient, it's dangerous. During early counter-UAS drills at domestic installations, researchers found that more than half of civilian bystanders misinterpreted defensive actions as offensive, assuming the military was "shooting at hobbyists." That perception gap is fertile ground for manipulation. When the public can't tell the difference between defense and aggression, adversaries don't need drones to cause division; they need perception.

The answer is not panic, it's participation. Fear immobilizes. Awareness empowers. Every local government, emergency management office, and civic organization must begin treating low-altitude security as part of community preparedness. Just as schools practice fire drills, towns should rehearse airspace incident protocols, not because war is imminent, but because safety should never depend on luck. When citizens understand their role in observation and reporting, they stop being spectators and become sentinels. Homeland defense isn't a spectator sport; it's a civic discipline.

During one counter-UAS workshop, an old sergeant looked at a swarm of consumer drones buzzing overhead and muttered, "That's not air superiority, that's air stupidity." Everyone laughed. But the room understood the point. Technology without accountability is noise with wings. Progress isn't the problem. Ignorance is.

In the next section, *"Bureaucracy's Blind Spots: How Agencies Collide While Drones Fly Free,"* we examine the policy confusion that hampers domestic drone response, from outdated FAA frameworks to turf battles between federal and state agencies. Because before the nation can defend its skies, it must first decide who's actually in charge of them.

Section 2: Bureaucracy's Blind Spots: How Agencies Collide While Drones Fly Free

In Washington, *the problem is never that no one is in charge; it's that everyone is.* That observation, offered by a retired DHS policy director, captures America's drone dilemma with surgical precision. The sky may be shared, but the system beneath it is anything but. Each agency treats the airspace like a turf war with invisible fences: the FAA governs flight, DHS handles domestic threats, DoD manages national defense, and local law enforcement is left to deal with everything that falls between. This patchwork worked well enough when airspace meant airplanes. It collapses when the same sky hosts tens of thousands of drones, most of which are smaller than a seagull and faster than any response protocol can track. The adversary doesn't need to outsmart the system; they need to fly through the seams where authority ends and coordination begins.

The FAA, despite its legacy and expertise, was established for safety, not security. Its mission is to promote aviation growth while maintaining safe operations, not to repel adversaries. It can issue flight restrictions, revoke licenses, and fine operators, but it has no operational arm to intercept or neutralize rogue drones. In effect, it's a referee in a game with no rulebook, surrounded by players who don't even know they're playing. Even when violations are detected, the process remains procedural: notifications, investigations, and administrative reviews. These steps are essential in peacetime but painfully slow in crisis. A drone doesn't wait for due process; it just flies on.

Meanwhile, DHS understands the threat better than anyone, but it lacks the legal authority to act decisively against airborne systems inside U.S. borders. Agencies like CISA and the Office of Intelligence and Analysis can detect vulnerabilities and issue advisories. Still, unless the threat escalates into terrorism, DHS must rely on local or federal partners for enforcement. The result is a paradox: the agency most aware of the danger is the one least empowered to respond. It's as if the fire marshal could see the flames but had to file a request before grabbing the hose.

The Department of Defense, by contrast, is the most capable actor in the sky, and the most constrained. The Posse Comitatus Act prohibits direct military action on U.S. soil except under specific circumstances. That means even if NORAD or NORTHCOM detects a drone near sensitive infrastructure, it cannot act without federal authorization and interagency coordination. The irony is bitter: the same command that tracks intercontinental missiles can't engage a drone hovering over a refinery without legal consultation. In a world where threats evolve by the minute, bureaucracy still measures time by the memo.

Local law enforcement sits closest to the threat but farthest from the tools to stop it. Most departments lack radar coverage, countermeasure training, and legal clarity. If a drone flies near a stadium or airport, officers can document and report it, but shooting it down could violate federal airspace laws. In many cases, the safest and only option is observation. But observation without authority isn't safety; it's helplessness. That helplessness fuels both public frustration and the confidence of adversaries.

Every major federal response exercise highlights the same flaw: a lack of coordination without apparent command authority. When multiple agencies arrive on scene, the question isn't what to do; it's who decides. A counter-UAS scenario at a critical infrastructure site might include local police, DHS analysts, FAA advisors, and a DoD liaison. Each brings expertise. None brings final authority. In a crisis, that translates to paralysis. Coordination works on paper. Command works in reality. The United States currently has plenty of the first and not enough of the second.

Congress, for its part, has tried, and failed, to untangle the mess. The Preventing Emerging Threats Act of 2018 granted limited counter-UAS authority to DHS and DOJ, but only for specific mission sets: federal facilities, border protection, and significant events. Renewal efforts in 2023 stalled over privacy concerns and partisan disputes. The result is a nation that is both partially protected and partially exposed, governed by laws that treat a 2025 drone as if it were still 2005 technology. Congress seeks perfect legislation in a world that moves at an imperfect pace. But perfection, delayed long enough, becomes negligence.

Even when policy allows action, agencies often hesitate, not out of fear of failure, but out of fear of blame. In the military, initiative is highly valued and rewarded. In bureaucracy, it's reviewed. No one wants to be the official who authorizes a countermeasure that damages private property or interrupts communications. So, agencies default to caution, which in practice becomes inaction. The safest career move is to do nothing until someone else makes a mistake. That mindset, left unchallenged, will cost more than reputation. It will cost readiness.

Consider a sold-out sports event. A drone appears above the crowd, carrying what looks like a camera, but it could be something else. The FAA is notified but cannot engage. The FBI is informed but lacks immediate jurisdiction. The local police can't risk shooting into a crowd. By the time decisions reach the top, the drone is gone. And the next one learns from the first. That's how adversaries exploit processes, not by overpowering systems, but by outpacing them.

During a joint-agency workshop, a colonel once joked, "If a drone crosses three jurisdictions, it earns frequent flyer miles before anyone can stop it." The room laughed because it was true. Humor in these settings isn't escapism; it's a coping mechanism for professionals who understand the absurdity of operating under laws written for an analog sky. Progress requires acknowledging absurdity, without succumbing to cynicism.

Reform begins with three steps. First, establish a unified national counter-UAS command network that combines airspace oversight, incident response, and intelligence analysis under a single joint directive. Second, empower state and local agencies with tiered engagement rules based on proximity, risk, and escalation thresholds. Third, mandate continuous legal modernization, requiring legislative reviews every two years to prevent decade-old laws from governing year-old tools. Accountability must evolve faster than technology. Otherwise, the law will always be one flight behind the threat.

In the next section, *"The New Minutemen: Citizens as the First Observers,"* the focus shifts from institutions to individuals, the citizens, employees, and communities who represent the nation's proper early warning system. Because until the bureaucracy can see clearly, the public must.

Section 3: The New Minutemen: Citizens as the First Observers

"Eternal vigilance is the price of liberty.", Wendell Phillips, 1852

Every generation produces its guardians. Some stood on village greens with muskets; others now stand in parking lots with smartphones. The tools have changed, but the principle remains: the defense of a free nation begins with those willing to pay attention. Wendell Phillips called it "eternal vigilance," and while the phrase has aged, the responsibility remains. In fact, it's never been more relevant. The slogan "See something, say something" may have been born in the shadow of 9/11, but it never matured into a doctrine. Most citizens still associate national security with distant agencies and faraway threats. Yet in the drone age, vigilance is no longer the government's monopoly. Security begins where people live, work, and worship. The modern minuteman doesn't wear a uniform; he carries a sense of responsibility.

Every person with a phone or tablet already participates in a distributed sensor grid, whether they realize it or not. Cameras, GPS, and network data form the backbone of situational awareness. In the right hands, this information can prevent catastrophe. During a 2022 wildfire in California, civilians utilized drone footage and crowd-sourced images to assist first responders in tracking the fire's movement in real-time. That same coordination, if properly directed, could identify drone incursions, track flight patterns, and record unusual aerial behavior near critical infrastructure. The power already exists; it only needs to be organized. Imagine a voluntary network that allows citizens to report airspace anomalies to a central clearinghouse, not to snoop, but to strengthen response. The technology is feasible. What's missing is a civic mindset.

Of course, the greatest danger in public participation is overreaction. Awareness must never become paranoia. Citizens should not view the sky with suspicion, but with discernment. Not every drone is a threat, and not every operator is an adversary. Vigilance means observation without hysteria, engagement without escalation. To achieve that balance, communities need education. Cities and counties could integrate short airspace awareness sessions into community policing events, emergency preparedness fairs, or even local school programs. If the goal is prevention, then education serves as a deterrent.

History shows that when local leaders act decisively, national strategy follows. Sheriffs, mayors, and county emergency managers are the connective tissue between federal institutions and the public. They don't need new authority; they need clarity. A local official who knows how to report a drone incident efficiently can cut hours off federal response time. Small towns have stopped wildfires, found fugitives, and coordinated rescues long before national resources arrived. The same principle applies to aerial threats. The first call doesn't have to come from Washington. It can come from a warehouse worker who notices something odd above the power substation at dawn. That's not fear; that's stewardship.

Civil vigilance, however, raises legitimate questions. Where is the line between awareness and intrusion? How do we prevent citizen reporting from turning into suspicion of neighbors or discrimination against hobbyists? The answer lies in intent. Awareness must always serve protection, not persecution. A well-informed citizen can recognize patterns, not people, behaviors, not backgrounds. The point isn't to label the operator; it's to identify unsafe or unlawful activity. Training civilians to think in terms of risk, not identity, ensures vigilance remains inclusive rather than accusatory.

The tools to enable this safely already exist. Apps that detect local drone frequencies, public mapping platforms that highlight restricted zones, and integrated emergency reporting channels could transform every community into a connected defense cell. Technology should empower citizens, not isolate them. One pilot program in Texas allowed residents to opt in to airspace alerts for their ZIP code. If a drone entered restricted airspace, the system sent notifications and instructions on how to report safely. The result was not chaos, but cooperation. Awareness, properly guided, creates calm.

The original American Minutemen were ordinary citizens, farmers, merchants, and laborers, who trained to respond within minutes of a threat. Their strength was not in their weapons, but in their readiness. The 21st-century equivalent is mental, not martial. It's the parent who reports unusual drone activity near a school, the city employee who notices repeat flyovers near a reservoir, or the jogger who spots a quadcopter hovering over a highway bridge. Each observation, logged and verified, strengthens collective defense. In this sense, citizens are not the last line of defense; they are the first.

In too many crises, the public has been cast as an audience, watching, waiting, and wondering when someone else will fix it. That attitude may have worked in the analog age, but it collapses in a connected one. If every citizen understood their role in safeguarding their community's digital and physical airspace, the collective deterrence effect would be enormous. Adversaries thrive in anonymity. The moment the public starts paying attention, their margin for success shrinks. An alert community is more complicated to infiltrate than a well-funded one.

For public vigilance to be effective, citizens must trust the institutions to which they report. When they don't, participation evaporates. If a resident believes their report will be ignored, they won't file it. If they fear their data will be misused, they won't share it. Trust, therefore, is not a courtesy; it's an operational necessity. Transparency in communication, explaining how information is used, what outcomes it drives, and where accountability lies, turns skepticism into cooperation. When citizens know they're not being exploited, they become invested.

In a recent workshop, a firefighter joked, "The best early warning system is still the nosy neighbor with a camera." He wasn't wrong. That neighbor, often dismissed as overinvolved, might be the one who notices the pattern everyone else overlooks. In national security, there's a fine line between nosiness and nobility. The difference is purpose.

Vigilance does not belong to one agency, party, or generation. It belongs to the republic. The lesson of every conflict, from the Revolution to the War on Terror, is that freedom endures when citizens remain vigilant. The new Minutemen don't carry muskets or powder; they bring awareness, context, and courage. Uniforms or drills won't measure their readiness; instead, it's their ability to recognize that peace is maintained through participation. The sky is not someone else's problem. It's everyone's responsibility.

In the next section, "*Lessons in Coordination: Bridging Civil and Federal Readiness*," we explore how to connect this grassroots vigilance to structured response, transforming awareness into capability and ensuring that when citizens sound the alarm, institutions actually hear it. Because defense without connection is noise, defense with coordination becomes a source of strength.

Section 4: Lessons in Coordination: Bridging Civil and Federal Readiness

"You can't protect what you don't coordinate.", Joint Interagency Task Force Training Note, 2022.

The United States does not suffer from a shortage of intelligence, resources, or capability; it suffers from a chronic lack of coherence. Every central homeland security review since 2001 has diagnosed the same condition: coordination collapses precisely where jurisdiction begins. Federal agencies collect data, state authorities enforce policy, and local responders handle the immediate scene. Yet, these components rarely align in real-time. By the time information travels from local observation to national response, the moment has already passed. Coordination should be reflexive, not a request. Until that becomes reality, readiness remains a theory dressed up as a plan.

Federal frameworks excel at strategy but tend to fracture under pressure. They operate under ideal conditions, characterized by clear communication, rigid chains of command, and designated authority. Local realities, however, are messier. A sheriff's department or small-town emergency manager doesn't have a drone liaison or access to national databases. When a suspicious drone appears near a refinery, the responding officer must navigate a maze of reporting protocols before anyone with decision authority even sees the alert. By then, the drone had vanished, leaving behind nothing but frustration and a trail of paperwork. Doctrine means little without agility. To bridge that gap, federal agencies must stop treating local responders as policy extensions and start treating them as execution partners.

Traditional command structures rely on verticality: information flows up, decisions come down. That model collapses under the velocity of drone threats. The solution lies in networked coordination, shared databases, open communication channels, and decentralized reporting. When field officers, analysts, and federal monitors see the same information simultaneously, response time drops from minutes to seconds. That shift requires more than technology; it demands a cultural overhaul. Agencies must learn to share credit as quickly as they share data. Coordination is not about who gets the headline; it's about who gets there first.

One reason coordination falters is the obsession with guarding "intelligence" as if it were a national treasure. Raw information, however, is not power until it's interpreted and shared. A private energy company might detect unusual drone activity over transmission lines but hesitate to report it, fearing regulatory scrutiny or public panic. Meanwhile, law enforcement may suspect sabotage but lack corroboration. Each side waits for confirmation that the other already has. This culture of caution wastes time and erodes trust. What's needed is less competition for information and more collaboration around interpretation. Intelligence is a product; defense is a process.

Joint exercises should expose weaknesses, not confirm comfort. Too often, agencies rehearse idealized scenarios designed to showcase readiness rather than reveal blind spots. A real exercise should simulate confusion, overlapping jurisdictions, delayed communication, and missing data. That friction is not failure; it's training. When agencies discover their blind spots during rehearsal, they prevent them during a crisis. In one interagency drill conducted in the Southwest, it took federal and local responders four hours to confirm they were tracking the same drone. The lesson wasn't humiliation; it was revelation. Readiness grows from discomfort, not display.

Coordination is not only cultural; it's technical. Systems often fail to communicate because they weren't designed to. FAA databases, law enforcement systems, and defense intelligence networks operate on incompatible architectures. A shared national counter-UAS data environment would bridge that gap, allowing real-time visibility across agencies. Disciplined integration eliminates the need for additional bureaucracy, and technology that functions as a translator enables coordination to occur routinely rather than in response.

Top-down coordination fails when it ignores the people closest to the threat. Local responders and community leaders have a deeper understanding of the terrain, behavior patterns, and emerging risks, often long before national analysts do. Thus, federal response teams should treat local expertise as intelligence, not anecdote. A deputy in Kansas who observes recurring drone activity near a grain elevator may lack the resources to analyze it, but he still holds critical context. Listening downward is as vital as reporting upward. Proper coordination is vertical, horizontal, and humble.

Private industry controls most of America's infrastructure, including power, water, energy, transportation, and data networks. That means the majority of potential targets lie outside direct government control. The public-private partnership model must evolve from a focus on contracts and compliance to one of continuous collaboration and mutual benefit. Industry leaders should have direct lines to security liaisons who can provide real-time advice, not just quarterly reports. Likewise, government agencies must view industry not as a risk to manage, but as an ally to empower. A cyberattack on a pipeline and a drone intrusion over it are two sides of the same vulnerability; both demand a unified response.

Coordination is not only a defensive necessity; it's a strategic deterrent. When adversaries see seamless communication between federal, state, and local entities, they lose confidence in ambiguity. Deterrence today is not about stockpiles or superiority; it's about unity under pressure. The appearance of confusion invites probing. The appearance of coordination prevents it. An adversary will always test what looks uncertain.

Coordination cannot be improvised. It must be trained as deliberately as marksmanship or cyber defense. Every federal training academy, from FEMA to DHS to the military, should include interagency simulation as a core competency. Future leaders must learn not just how to command, but how to connect. A commander who can't coordinate is just a spectator with authority.

At a homeland security conference, a county emergency manager remarked, "We've got more acronyms than aircraft." The audience laughed, then nodded. It was a fair observation; coordination dies in the face of jargon. If no one outside your agency understands your language, then you're not securing anything; you're just speaking to yourself. Explicit language is the first act of coordination.

The future of domestic security will depend less on new technology and more on synchronization of existing capacity. The systems, expertise, and personnel already exist. What's missing is shared rhythm, the steady cadence that transforms separate efforts into a unified response. That rhythm must be practiced, institutionalized, and expected. Coordination is not an event; it's a culture. The nation that learns to act together will always outpace the one that merely reacts separately.

In the next section, *"The Shared Responsibility Doctrine: Building a Culture of Preparedness,"* we'll conclude this chapter by weaving together every layer, government, industry, and the public, into one principle: defense is not delegation; it's participation. Because in a world where anyone can launch a threat, everyone must be part of the solution.

Section 5: The Shared Responsibility Doctrine: Building a Culture of Preparedness

"Security is not a product delivered by government; it is a practice shared by citizens.", Adapted from a 2022 Homeland Security Seminar.

Every generation redefines what defense means. For the Founders, it meant raising militias and guarding borders. For the Cold War generation, it meant deterrence through strength. In the twenty-first century, defense is no longer measured by the size of the military but by the strength of the nation's collective awareness. The Shared Responsibility Doctrine emphasizes that security requires active participation and ongoing effort from the entire community, rather than relying solely on government delivery. Preparedness begins when citizens stop seeing themselves as passive beneficiaries of safety and start taking an active role in it.

For too long, public safety has resembled a service contract, government as provider, people as consumers. That model has expired. The complexity of modern threats outpaces the capacity of any single institution. Consequently, preparedness must evolve from a state of dependency to one of partnership. Citizens must trust that the government will respond effectively; the government must trust that citizens will act responsibly. That trust cannot be legislated; it must be earned through transparency, communication, and shared accountability. Every citizen who learns to identify, report, or prevent threats extends the reach of national defense. That's not idealism; it's arithmetic.

Every few years, a new awareness initiative appears, a slogan, a brochure, a website. Then it fades when the news cycle moves on. What we need isn't another campaign; it's a culture. Culture is what people do without being told. It's what communities practice when no one is watching. Preparedness must become a habit woven into everyday life, much like driving defensively, maintaining smoke alarms, and locking our doors. If every citizen viewed airspace awareness and infrastructure protection as a civic norm, the deterrent effect alone would reshape the national threat landscape. You don't need a battalion when you have a neighborhood that pays attention.

Thus, national security should begin with local preparedness. The simplest way to strengthen the homeland is to decentralize resilience, to empower counties, towns, and even neighborhoods to take ownership of their security posture. Imagine annual readiness drills hosted by local governments, integrating first responders, volunteers, and small businesses. Imagine school curricula teaching not fear, but awareness: how drones, AI, and cyber tools can both help and harm society. When readiness becomes local, it becomes personal. And when it becomes personal, it becomes sustainable.

Private industry now holds more influence over national resilience than ever before. From data centers to drone manufacturers, corporate decisions shape the integrity of America's digital and physical infrastructure. The Shared Responsibility Doctrine requires companies to view security not as a matter of compliance, but as a matter of conscience. That means responsible engineering, transparency in data practices, and a refusal to cut ethical corners for the sake of efficiency. When companies safeguard their systems, they protect their customers, and by extension, the nation. Patriotism can be practiced through production as much as through policy.

For the government, the doctrine requires humility and a focused approach. Bureaucracy must stop confusing control with leadership. The goal is not to dominate the public's role in defense, but to empower it. Federal agencies should streamline their reporting systems, enhance communication with local responders, and provide the necessary resources to enable local responders to be self-sufficient. The more the government tries to centralize security, the slower it becomes. The more it decentralizes awareness, the stronger the network grows. Empowerment is not loss of authority; it is the multiplication of capability.

Resilience begins at home. Families that teach responsibility, situational awareness, and self-discipline produce citizens who contribute to stability rather than chaos. Churches, community centers, and civic groups can reinforce these values without politicizing them. Preparedness is not a partisan virtue; it is a moral one. Faith-based organizations have historically been the first to respond to disasters and crises. They can also be the first to educate, mobilize, and sustain communities before the next emergency arises. A prepared nation begins with prepared households.

Preparedness must not breed paranoia. The objective is calm readiness, not fear-driven reaction. Training should emphasize observation, communication, and composure under stress. Communities that practice together build confidence; confidence prevents panic. Preparedness does not mean expecting the worst; it means being ready for it. There's a difference between vigilance and hysteria, and it's called training.

One homeland security instructor begins every session by saying, "You're not paranoid if you're paying attention; you're just a responsible adult with Wi-Fi." The room always laughs, and then it listens. Humor can help disarm anxiety and make it easier for people to absorb harsh truths. Awareness need not be somber; it can be empowering, even hopeful. A nation that can laugh together can coordinate together. Sarcasm, when used well, is just patriotism with a sharper edge.

To move from doctrine to action, we need a blueprint. First, educate continuously by integrating awareness of airspace, cyber, and infrastructure into public education and professional training programs. Second, communicate transparently: ensure that agencies, industries, and communities share real-time information without excessive red tape. Third, conduct joint exercises regularly, including national and regional readiness exercises that involve civilians, local responders, and federal oversight. Fourth, incentivize participation by offering recognition and support for civic engagement in security initiatives, from youth programs to neighborhood networks. Fifth, sustain momentum: treat preparedness as a standing priority, not a seasonal campaign. A culture of preparedness is not created by command; it's cultivated by continuity.

Preparedness is not simply about safety; it's about stewardship. A secure nation is not defined by how it reacts to a crisis, but by how it anticipates and prepares for it. In Scripture, Proverbs 22:3 reminds us: "The prudent see danger and take refuge, but the simple keep going and pay the penalty." That verse could serve as the unofficial creed of homeland defense. Prudence is not fear; it's foresight. A vigilant nation honors both its freedom and its faith by preparing for what it hopes will never come to pass.

The Shared Responsibility Doctrine can be summarized in five words: If everyone guards something, nothing falls. Each citizen, agency, company, and leader must see themselves as part of one living network, interdependent, informed, and invested in the same goal. The defense of the homeland does not begin with radar or legislation; it starts with ownership. When people take responsibility for their corner of the republic, the nation becomes unbreakable.

In the next chapter, "Eyes Everywhere: Surveillance and the Vanishing Privacy Line," we leave the frontline of physical defense and step into the quieter, more insidious battlespace, the one fought through observation, data, and consent. Drones no longer just occupy airspace; they occupy attention. The same technologies designed to protect us have begun to monitor us, record our actions, and, over time, influence our behavior. As boundaries blur between public safety and personal privacy, between transparency and intrusion, the question is no longer whether we are being watched, but by whom, for what purpose, and with what consequence. Because the sky isn't our only exposure; in the age of constant visibility, the real frontier of defense is the self.

Chapter 5 – Eyes Everywhere: Surveillance and the Vanishing Privacy Line

Section 1: Watching the Watchers: From Satellites to Street Corners

"Quis custodiet ipsos custodes?"

Juvenal

"Who will watch the watchers?"

Mrs. Alvarez wasn't paranoid. She simply paid attention. Her neighbor's new drone had a habit, hovering near her backyard every afternoon, right when her daughter practiced gymnastics. At first, she dismissed it. Maybe it was watching birds. But the drone returned. Same time. Same angle. Every day. She reported it. The police shrugged. "Nothing illegal," they said. "It's private airspace above your fence line." So, she planted trees. Built a canopy. And wondered, quietly, when did our backyards stop being ours?

Once upon a time, surveillance was obvious. Cameras were bulky, satellites were secret, and suspicion had a sound, the whir of film reels or the static of intercepted radio chatter. Today, surveillance is silent, weightless, and everywhere. It floats not on vapor, but on data. The irony is hard to miss. The same society that once feared "Big Brother" now willingly carries him in its pocket. The line between watcher and watched has blurred so thoroughly that most people no longer notice which side they're on. Every time someone unlocks a phone with a face scan, checks a GPS route, or uploads a selfie from the backyard, they contribute another tile to an ever-growing digital mosaic, a self-portrait painted by algorithms. The eye in the sky has multiplied, and now it lives among us.

The story of modern surveillance began with noble intent: defense, deterrence, and national security. Early satellites weren't built to win wars; they were built to prevent them. They watched the skies for ballistic missiles and troop movements, turning surprise into data. But as the technology matured, it followed a predictable American trajectory, from classified to commercial, from exclusive to accessible. What was once the privilege of intelligence agencies became the pastime of teenagers with drones. Surveillance democratized. Observation became recreation. Today, a consumer with a $600 drone can achieve a level of aerial awareness that would have cost governments millions during the Cold War. That's progress, and peril. When surveillance becomes entertainment, privacy becomes collateral.

Most Americans are unaware of the number of eyes that follow them daily. In major cities, closed-circuit cameras line the streets. Retail stores track movement through AI analytics. License plate readers record the path of every car. Every purchase, click, and commute feeds into invisible ledgers owned by corporations and government entities alike. None of this happened overnight. It occurred in small, reasonable steps, each justified by safety, convenience, or cost savings. A new camera after a crime wave. A tracking app for traffic management. A social media filter that maps your face for "fun." The cumulative effect is omnipresence disguised as progress. Surveillance didn't break down our doors; it walked through them politely, offering free Wi-Fi.

Human psychology craves safety, even at the cost of solitude. The idea that "someone is watching" provides reassurance until it becomes routine. The surveillance state doesn't need to hide; it simply needs to feel normal. People accept what protects them, and the bargain is always the same: we'll keep you safe if you let us watch. Slowly, a culture of casual consent emerges. Cameras don't just record; they reassure. "Someone's watching," we tell ourselves. "That means someone cares." But what happens when care turns to control, when the same data used to protect can also predict, or punish? That's the dilemma of the modern age: safety as seduction.

Military professionals often speak of "situational awareness", knowing one's environment, understanding one's field of view, and anticipating what comes next. That principle now applies to civilians, but the awareness runs both ways. When the watcher can also be watched, control becomes illusion. The drone operator's camera may capture footage, but network beacons and satellite signatures track that same drone. Every observation leaves its own trail. The age of one-sided surveillance is over; this is the era of mutual visibility.

Even governments now live under scrutiny. Citizens livestream protests, upload evidence, and track official actions in real time. The same lens that enforces control can also expose corruption. Surveillance, in this light, is both a threat and a tool, capable of protecting liberty or extinguishing it, depending on who holds the controls.

The danger isn't just being seen; it's being remembered. In previous generations, privacy returned with time. What you said, did, or wrote faded into the past. Today, data never dies. Every keystroke, image, and location check becomes part of a permanent, searchable history, not just of individuals, but of entire populations. The result is subtle but profound: the erosion of forgetfulness. Society has lost its ability to move on. People are no longer judged by who they are, but by the digital footprints they've already left behind. Society rarely allows genuine redemption when every past action remains permanently recorded.

Most people think they control their privacy through settings, consent boxes, or "agree" buttons. They don't. Those mechanisms offer the illusion of control while granting permission for precisely what users think they're limiting. Opting out of surveillance today is like trying to stay dry in a thunderstorm by holding an umbrella over one drop. The system is too big, too interconnected, too profitable to stop voluntarily.

Control has become choreography, structured motion within boundaries already decided by others. The best illusion is the one that makes you feel free while you follow the path that has been designed for you.

A nation that forgets privacy forgets humility. There is a moral dimension to being unseen; it preserves dignity, fosters self-reflection, and promotes grace. Faith teaches that accountability to God precedes accountability to government. But in a world of cameras and databases, that order begins to blur. When everything is observed, genuine introspection becomes performance. People start behaving not according to conscience, but according to expectation. Freedom erodes not through chains, but through observation.

During a training seminar, a senior analyst once quipped, "If you want to hide something from the government, write it down by hand; no one under thirty will ever find it." The laughter in the room said as much about truth as it did about irony. Technology has given us endless access, but very little privacy. What was once sacred, the letter, the conversation, the unrecorded thought, now feels like an endangered species.

In the next section, "*The New Panopticon: How Convenience Turned to Complacent Consent*," we'll explore how society's appetite for comfort and connectivity gradually replaced skepticism with surrender, how people traded privacy for practicality, and how the watchers became invisible not by stealth, but by invitation. Because in the modern world, we're not losing privacy by force; we're giving it away, one click at a time.

Section 2: The New Panopticon: How Convenience Turned to Complacent Consent

"The more comfortable we become, the less we question who made it so." Adapted from Aldous Huxley.

It didn't happen through tyranny; it happened through terms of service. No force, no threat, no coup, just a quiet agreement we clicked before scrolling down. The new surveillance state wasn't built with barbed wire or bunkers; it was constructed with software updates and subscription models. What once required coercion now operates on convenience. The New Panopticon creates a digital cage that disguises control as comfort, presents visibility as a voluntary choice, and embeds control within a user-friendly design. The brilliance of the system is that it doesn't need to hide; it only needs to make you feel safe.

When social media first emerged, it promised to connect people. People didn't sign up to be watched; they signed up to belong. The reward was instant validation, and the price was invisible. Now, billions of users voluntarily share their locations, relationships, and emotions in real-time. They photograph their own routines and tag them for public consumption. In any other century, this would have looked like madness; today, it looks like Tuesday. What started as curiosity has become codependence, a psychological trade between attention and affirmation. Every app that tracks your preferences also teaches you what to prefer. That isn't influence; it's conditioning.

Jeremy Bentham's original Panopticon was a prison design that allowed guards to observe inmates without being seen. The genius of the model wasn't total surveillance; it was uncertainty. Prisoners behaved because they might be watched at any time. Modern society has recreated that model voluntarily. We behave as though every act were visible, every statement were recorded, every choice were remembered. We self-censor to avoid conflict, tailor our opinions to algorithms, and edit ourselves in anticipation of being judged. This control system achieves maximum efficiency by relying on active participation instead of traditional guards. The camera doesn't have to move if the subject polices himself.

Every major innovation in personal technology has arrived with the same promise: simplicity, connection, ease. Each one made life smoother; each one made privacy thinner. The smart speaker that listens to commands also listens to conversations. The security doorbell that keeps burglars out also records neighbors passing by. The smartwatch that tracks heart rate also tracks location. These devices are marketed as empowerment, yet they function as sensors in a global web of perpetual observation. Convenience is the camouflage of control. And like any camouflage, it works best when people want to believe in the scenery.

The New Panopticon doesn't operate on ideology; it runs on commerce. Data is the new currency, traded, analyzed, and repackaged faster than any government could legislate it. Every digital interaction generates information that feeds predictive algorithms. These algorithms then refine how markets, media, and even political campaigns target individuals. You are not the customer; you are the product. The genius of this economy is that it doesn't demand your compliance; it earns it. The system rewards participation with convenience and punishes withdrawal with inconvenience. Opting out feels like opting backward. And so, even those who recognize the cost continue to pay it, with clicks, likes, and loyalty points.

Once upon a time, Americans instinctively distrusted authority. They questioned power, challenged intrusion, and demanded accountability. Now, skepticism has been replaced by fatigue. People are too busy to fight every policy update or privacy breach. Outrage has become background noise. When surveillance becomes routine, resistance becomes rare. And when resistance becomes rare, democracy weakens, not through suppression, but through sedation. Complacency, not censorship, is the most reliable silencer of all.

The New Panopticon isn't a one-way mirror; it's a hall of mirrors. People not only accept surveillance, but they also imitate it. Every smartphone owner is now a potential recorder, every interaction a possible broadcast. We no longer distinguish between being watched and watching ourselves. Social media has transformed self-surveillance into identity performance; we curate our own observations, filter out our flaws, and call it authenticity. We have become both the subject and the guard. This isn't oppression; it's obsession.

Faith once taught patience, privacy, and restraint. Modern life teaches immediacy. The spiritual rhythm of reflection, listening before speaking, thinking before acting, has been replaced by the algorithmic rhythm of reaction. In a world where everything is captured, contemplation feels inefficient. That's why silence, solitude, and faith have become the last refuges of privacy. To pray is to unplug, to seek a channel no network can intercept. Faith offers what technology cannot: the assurance that some thoughts are still sacred.

During a cyber-ethics seminar, one instructor joked, "If the Founding Fathers had smartphones, we'd still be waiting for the Declaration of Independence to finish syncing." The room laughed, but the point stuck. Speed has replaced reflection. We no longer pause long enough to question what we've connected to. And while the convenience feels like progress, it also feels like drift, movement without direction. Freedom, after all, is not just the ability to act; it's the ability to refrain.

Breaking free from complacent consent doesn't mean abandoning technology; it means mastering it. The challenge is to restore awareness, to recognize that privacy and progress are not opposites. Governments must enforce boundaries on data collection. Corporations must choose ethics over efficiency. Citizens must relearn the habit of curiosity, not the curiosity that clicks, but the curiosity that questions. Liberty doesn't disappear in an instant; it erodes in increments. Therefore, the task of every generation is to notice when the increments start adding up.

In the next section, "Data Is the Payload: Metadata and Predictive Policing," we move deeper into the machinery of modern surveillance, examining how invisible data, not visible footage, has become the actual instrument of control. Because in the digital age, the weapon isn't the camera; it's the code behind it.

Section 3: Data Is the Payload: Metadata and Predictive Policing

"We have never been more connected, nor more categorized.", Data Ethics Symposium, 2023

Not all payloads explode. Some simply accumulate. In the drone era, we imagine threats as physical, machines, propellers, or payloads of steel or plastique. But the most potent payloads today are invisible: streams of metadata, harvested, sorted, and weaponized by algorithms. Metadata is information about information, the who, where, when, and how of every digital act. It's not the message you send, but the record that you sent it; not the video you film, but the geolocation and timestamp behind it. Individually, these data points seem harmless. Combined, they create a portrait sharper than any photograph, one that knows you better than you know yourself. The modern battlefield isn't measured in miles; it's measured in metadata.

People take comfort in the idea that their data is "anonymous." It isn't. Anonymized data is simply pseudonymous, stripped of names but not of identity. Algorithms reassemble the fragments in seconds, connecting dots between purchase patterns, movement history, and online behavior. A phone that moves from a home to a workplace to a gym doesn't need a name tag to be identified. Its routine gives it away. Privacy, once defined by secrecy, is now undermined by predictability.

Law enforcement agencies worldwide are utilizing predictive analytics to anticipate crime before it occurs. The idea is appealing: use data to deploy resources efficiently, prevent danger, and save lives. But predictive policing carries a moral cost; it transforms probabilities into presumptions. Algorithms trained on historical data inevitably reproduce the biases embedded in that data. For example, neighborhoods that were previously over-policed are flagged again, perpetuating cycles of scrutiny without cause. The code may be neutral; the inputs are not. And unlike a human officer, an algorithm cannot explain its judgment; it only executes it. In the wrong hands, predictive policing becomes preemptive profiling.

The genius of metadata is subtlety. It doesn't need to look intrusive to be intrusive. A drone hovering overhead might draw suspicion; an app collecting motion data does not. A camera mounted on a pole feels like surveillance; a phone in your pocket feels like freedom. Yet both feed the same machine. When surveillance hides in routine, resistance fades into resignation. The result is a form of control so efficient it feels voluntary.

While governments use data to predict threats, corporations use it to predict behavior. The difference lies in motive; the method remains identical. Every click, search, and pause tells a story, not about what we did, but about what we'll do next. Corporations trade on anticipation, packaging patterns into profit. Over time, predictive algorithms stop reflecting behavior and begin to direct it. You're not just being watched; you're being guided, nudged toward the purchase, the vote, the opinion most compatible with your profile. Freedom that can be forecast is freedom already influenced.

The Department of Defense and Homeland Security increasingly relies on algorithmic analytics for threat detection, flight monitoring, and border management. These systems enhance speed and scope, but they also raise existential questions. When machines start making decisions faster than humans can review them, accountability becomes theoretical. Who bears responsibility for an automated error: the coder, the operator, or the command authority that approved the process? As autonomy grows, so does opacity. In warfare, this creates operational advantage. In governance, it generates moral risk. A republic built on human judgment must decide how much of its conscience it's willing to outsource to code.

The more data collected, the less clarity achieved. Information without interpretation becomes noise, and the temptation to automate interpretation creates dependency. Analysts now spend more time managing systems than understanding the people those systems are designed to monitor. Technology has inverted the intelligence hierarchy; machines analyze, humans approve. But machines don't understand context, humor, or grace. They can recognize patterns, but not repentance. When data becomes the lens through which we see humanity, we risk mistaking precision for wisdom.

Faith traditions have long warned against the illusion of omniscience. Only God sees all, and even then, He judges with mercy. Machines see all but forgive nothing. That distinction is crucial. When human systems assume divine vision without divine compassion, justice becomes arithmetic. A data-driven society must reintroduce morality into its models, not through slogans, but through restraint. Every algorithm should be tested not just for accuracy, but for equity. Every surveillance program should answer a moral question before a technical one: Does this protect people, or merely control them? Accountability is not a metric; it's a mandate.

During a defense technology summit, a colonel remarked, "We have so many analytics dashboards, I need a dashboard just to track my dashboards." The laughter that followed was knowing, not amused. Data promised clarity; instead, it created clutter. In pursuit of insight, we risk being overwhelmed by precision. Sometimes, the most accurate picture is also the least human.

Data itself isn't the enemy; misuse is. The challenge is not to reject technology but to govern it with conscience, to treat information as stewardship, not sovereignty. To achieve this, three principles must guide future policy. First, transparency: citizens should know what data is collected, who uses it, and for what purpose. Second, accountability: every algorithm that affects public life must be auditable through independent review. Third, proportionality: collection must be justified by necessity, not curiosity. Information is power, but only when handled responsibly does it remain righteous.

In the next section, *"Faith and Freedom: Balancing Transparency and Trust,"* we examine how moral conviction and civic values can coexist with surveillance, how faith-based ethics, constitutional principles, and cultural discipline must guide the use of data and technology before they begin to shape our lives. Because the question isn't just what we see; it's what we become when we stop looking inward.

Section 4: Faith and Freedom: Balancing Transparency and Trust

"Where the Spirit of the Lord is, there is liberty." 2 Corinthians 3:17

Technology doesn't just shape society, it tests it. Every generation encounters an invention that forces a reckoning: will this new power serve humanity, or will it begin to dominate it? For ours, that invention is data. We've built systems that kings could only dream of, tools that see nearly everything, know almost everyone, and predict roughly anything. But with that power comes a question no algorithm can answer: Should we?

Freedom begins not with capability, but with restraint. Some tools should remain unused simply because conscience says so. When nations lose that restraint, power becomes a form of appetite. And appetite, once normalized, rarely returns to moderation. The challenge isn't whether we can surveil more, it's whether we should.

Because modern governance often elevates transparency as a virtue, but rarely defines it. The idea sounds noble: open exchange, accountability through visibility. But transparency without trust breeds fear, not confidence. If every action, thought, and communication is exposed, the result isn't honesty, it's performance. People behave not as they are, but as they wish to appear. True integrity requires privacy: the freedom to think, to doubt, to grow without constant observation. When everything becomes public, authenticity becomes theater.

Long before the word "cyber" existed, the concept of internal discipline offered the first form of security; for example, conscience, not code, governed behavior. A person who believes in accountability, whether to principle, community, or law, doesn't need to be watched to act ethically. That is the essence of liberty: self-governance grounded in

119

moral conviction. In this sense, privacy and personal responsibility are twin guardians of freedom. One protects the space to act freely; the other ensures that freedom is used wisely.

Unchecked observation doesn't just erode privacy; it erodes purpose. When people begin living for the lens, they stop living for meaning. Surveillance teaches a reactive response, the impulse to respond to what others see, rather than reflection. That's why an over-surveilled society loses its sense of introspection and gains a culture of shame. Shame doesn't invite growth; it invites concealment. The danger of constant visibility isn't exposure, it's the loss of grace.

Governments justify surveillance using the language of protection, claiming it is necessary to detect threats, prevent attacks, and maintain order. Citizens accept it out of fear of terrorism, crime, or chaos. But a nation ruled by fear cannot remain free. Fear centralizes power; trust distributes it. Freedom requires that trust, in neighbors, in institutions, and in the rule of law over the rule of software. When trust dies, people demand control. When control arrives, freedom departs quietly.

To know everything is to risk judging everything. Every agency, company, and individual with access to surveillance data bears the burden of discernment. The temptation to misuse information grows with every byte collected. That's why transparency must always be paired with humility, the recognition that knowledge without wisdom is simply precision without purpose. A government that sees too much begins to assume it knows too much. That's how good intentions harden into overreach.

Privacy is not secrecy; it's a sanctuary. A home, a conversation, a moment of solitude, these are spaces where conscience breathes. They are not hiding places for wrongdoing; they are proving grounds for virtue. Without privacy, democracy cannot function effectively, and individuals cannot develop their character. That is why both the Constitution and civic tradition defend the unseen, not to shield guilt, but to preserve grace.

Security does not require omnipresence. It requires discernment. The role of surveillance should be protection, not possession. The goal is to prevent harm, not to dominate behavior. When government surveillance honors that distinction, it earns legitimacy. When it crosses it, it forfeits trust. Because the measure of a free nation is not how much it knows about its citizens, but how much it respects them.

During a privacy briefing, a data analyst once joked, "We're not Big Brother, we're just his helpful cousin with better Wi-Fi." The audience laughed, but the irony was evident. Surveillance doesn't announce itself as tyranny; it markets itself as service. That's why discernment is not paranoia; it's patriotism. A wise society laughs at irony but acts on truth.

Freedom demands vigilance, but of different kinds. Civic vigilance watches over the state; personal vigilance watches over the self. When both operate in tandem, society achieves balance. When either fails, the other falters. The responsible citizen doesn't reject technology; they humanize it. They demand that systems serve humanity rather than replace it. In an age of algorithms, reflection is a form of rebellion, a private act of autonomy that no data set can decode.

The goal is not perfect transparency or total privacy, but balance, a moral equilibrium between accountability and autonomy. That balance begins when people remember that liberty is not the absence of oversight, but the presence of self-restraint. Freedom cannot survive without trust, and trust cannot survive without humility. When humility replaces control, transparency becomes a virtue rather than a weapon.

This balance must be built into the architecture of governance. Surveillance programs should be subject to independent review, not just internal audits. Citizens should have access to clear explanations of what data is collected, how it's used, and what rights they retain. Agencies must distinguish between what is technically possible and what is ethically permissible. The question should never be "Can we?" but "Should we?"

Technology must be governed by principle, not just policy. That means embedding ethical review into every stage of development, from design to deployment. It means training engineers, analysts, and policymakers to think in terms of human dignity rather than just system efficiency. It means recognizing that the right to be unseen is not a relic of the past, it's a requirement for the future.

Civic education must also evolve. Citizens need to understand not only their rights but also their responsibilities in a society that is increasingly reliant on surveillance. This includes knowing how to report misuse, protecting their own data, and participating in public oversight. A democracy that delegates all vigilance to institutions will eventually lose its ability to self-correct. The public must remain engaged, not just as consumers of security, but as stewards of liberty.

Drones amplify these concerns significantly. Aerial surveillance changes the calculus of privacy. It's no longer just about what happens online; it's about what happens overhead. The legal frameworks built for letters and landlines are straining under the weight of satellites, smartphones, and autonomous aircraft. The right to be left alone must now include the right to be unobserved from above.

In the next section, "*The Right to Be Unseen: Revisiting Constitutional Privacy in the Drone Era*," we examine how those legal foundations must adapt. Because freedom was never meant to be invisible, but it was meant to be unwatched.

Section 5: The Right to Be Unseen: Revisiting Constitutional Privacy in the Drone Era

"The makers of our Constitution… conferred, as against the Government, the right to be let alone, the most comprehensive of rights, and the right most valued by civilized men.", Justice Louis Brandeis, 1928

When Justice Louis Brandeis described privacy as "the right most valued by civilized men," the world was still made of wires and paper. Trespass had a threshold. The government could violate privacy, but only by crossing a physical boundary. Today, that threshold is digital, silent, invisible, and omnipresent. Surveillance no longer knocks; it hovers, listens, and logs.

The founders never imagined a society where the government could see without entering, record without a warrant, and identify without a name. Yet that is the world in which modern Americans live. The Constitution protects "persons, houses, papers, and effects," but in an era when personal life is stored on servers and surveillance flies miles overhead, those boundaries are blurring. The right to be unseen, to exist without observation, has become the rarest liberty of all.

The Fourth Amendment was written to protect against unreasonable searches and seizures. But when data replaces property, what exactly is being seized? Courts have wrestled with this question for decades. In Katz v. United States (1967), the Supreme Court declared that "the Fourth Amendment protects people, not places." That principle became the foundation for modern privacy law, yet it was conceived long before the advent of drones, smartphones, and facial recognition.

Today's surveillance doesn't require a warrant to be conducted. It doesn't need probable cause to analyze. It simply exists, always on, always collecting. The result is an inversion of logic: privacy must now be justified; surveillance is assumed to be the norm. The burden has shifted from the state to the citizen. And in that shift, liberty begins to erode.

Drones further complicate the Fourth Amendment. Airspace used to be a clear legal domain, above ground, beyond reach. However, as drones operate below 400 feet, they occupy a gray zone that is neither public nor private, straddling both the sky and property. Can the air above a backyard be searched without consent? Is a drone-mounted camera equivalent to a police helicopter?

Courts have offered conflicting answers. In California v. Ciraolo (1986), aerial observation of a backyard from an altitude of 1,000 feet did not require a warrant. In Florida v. Riley (1989), helicopter surveillance conducted at an altitude of 400 feet was deemed lawful. Yet today, a drone hovering at the same altitude feels far more invasive, because it lingers, records, and repeats its actions. The law hasn't adjusted to scale or intimacy. It's trying to apply altitude to ethics.

The most remarkable legal irony of the drone age is that individuals and corporations now wield surveillance power once reserved for states. Private citizens can record, map, and monitor with minimal restriction, while governments remain bound by constitutional limits. The imbalance creates absurd scenarios: a homeowner can record a police officer, but a police officer may need a warrant to record the same homeowner.

This asymmetry isn't just inconvenient; it's dangerous. It blurs accountability, pits rights against responsibilities, and erodes the mutual trust that democratic governance depends upon. When privacy becomes a privilege rather than a principle, equality under the law collapses.

Corporate surveillance further complicates the equation. When users agree to the terms of service, they often surrender more privacy than the Fourth Amendment ever intended to allow. Legally, that consent is voluntary; practically, it's coercive. Modern life depends on platforms that demand compliance as a condition of participation.

The Constitution limits government intrusion, but it says nothing about corporations collecting data for profit. The result is an unguarded flank, private surveillance operating where public accountability cannot reach. When liberty becomes a mere formality, democracy becomes a mere formality.

Drones expose the hypocrisy of privacy law. Americans tolerate constant surveillance in digital space but recoil when they see it in the sky. That reaction reveals something profound: visibility still matters. What people can see reminds them of what's being taken. A buzzing drone above a backyard feels like trespass; an invisible server collecting browsing history does not. The law protects against the former and ignores the latter.

The right to be unseen isn't just physical, it's psychological. It's the freedom to exist without explanation. It's the ability to move, speak, and think without the expectation of being recorded. That freedom is foundational, not because it hides wrongdoing, but because it preserves dignity.

No responsible government can ignore security threats; no free nation can survive without limits on authority. The tension between safety and liberty is as old as the Republic itself. The challenge is not to eliminate surveillance but to discipline it. Drones can save lives during disasters, locate missing persons, and monitor wildfires. These uses serve the common good. But without oversight, the same systems can suppress dissent, chill speech, or intimidate citizens into silence.

And so, a free people must draw clear lines between protection and possession. Surveillance must serve the public, not subdue it. The goal is not omniscience, but proportionality.

Privacy is not a relic of the past; it is a test of modern principles. It asks whether a nation still believes in the individual's right to solitude, discretion, and autonomy. It asks whether law can keep pace with life, whether the Constitution can still protect what it cannot physically touch.

The framers grounded liberty in the belief that rights are inherent, not granted by courts, but recognized by them. That means privacy, too, is sacred before it is statutory. When the government begins to treat visibility as an entitlement, it confuses authority with authorship, forgetting that it was created by the people, not the other way around. Law should serve liberty, not redefine it.

During a privacy symposium, one attorney quipped, "The Constitution wasn't written with drones in mind, but it did a pretty good job without them." The room laughed because it was true. The principles remain sound; it's the policy that hasn't evolved. The Founders built fences around power. Modern technology keeps building ladders.

Therefore, protecting privacy in the drone era will require more than new laws; it will require a new philosophy. The nation must decide whether freedom means invisibility, anonymity, or simply the right to choose when to be seen. Congress must clarify ownership of low-altitude airspace. Courts must modernize interpretations of "search" and "seizure." Agencies must develop protocols that distinguish between surveillance for safety and surveillance for control.

However, more than that, citizens must reclaim the cultural value of discretion, the idea that not everything needs to be shared, recorded, or archived. The right to be unseen is not retreat; it is preservation. It is the space where conscience breathes, where thought matures, where liberty lives without performance.

When America learns to see privacy not as isolation, but as integrity, it will rediscover what Brandeis called "the right most valued by civilized men." That right is not a luxury; it is a boundary. And boundaries, once erased, rarely redraw themselves.

The consequences of this uneven regulatory terrain ripple far beyond the FAA. From local law enforcement to tribal leaders, the absence of standardized policies creates a vacuum that is quickly filled by guesswork, conflict, or inaction.

But fragmentation isn't just a failure of coordination, it's a failure of courage. In the absence of a unified plan, each stakeholder is left to either improvise or abandon their responsibilities.

In the next chapter, *"The Great Policy Gap: When Law Lags Behind Lift,"* we confront the legal turbulence that defines the drone era, how outdated regulations, fragmented authority, and political hesitation have left the nation vulnerable to both misuse and misunderstanding.

Because technology advances faster than bureaucracy, and if law fails to keep pace, freedom itself begins to fall.

Chapter 6 – The Great Policy Gap: When Law Lags Behind Lift

Section 1: The Legal Altitude Problem: Defining Airspace for the People

"The law must learn to fly as high as the machines it governs.", Adapted from a FAA Policy Brief, 2021

The city of New Orleans submitted its paperwork 42 days before the deadline. The request was simple: a temporary no-fly zone over the French Quarter during Mardi Gras. The floats were ready, the crowds were massive, and the drones were already airborne. The FAA's approval arrived, 11 hours too late. By then, three drones had buzzed parade routes. One narrowly missed a mounted police officer. Local officials held a press conference. The federal response: "We'll evaluate our process." The regional response: "Too little, too late." In a high-speed world, bureaucracy still operates at a pace reminiscent of the past.

The most valuable real estate in America is no longer the ground; it's the air just above it. Yet this space, the thin slice of sky between 0 and

400 feet, remains one of the most legally ambiguous domains in modern history. For centuries, property rights were clear: you owned the land and everything above it "to the heavens." That principle, *cujus est solum ejus est usque ad coelum*, worked fine until humanity learned to fly. Once airplanes entered the equation, Congress and the courts were forced to rethink ownership of the sky. The result was a patchwork of rulings, compromises, and outdated assumptions that still haunt policy today.

In a world where drones operate below traditional aviation corridors but above private property, the law struggles to govern altitude with clarity and precision. And altitude without clarity becomes chaos.

The FAA claims sovereignty over all navigable airspace in the United States. Historically, that began at 500 feet for fixed-wing aircraft; below that altitude, the line blurred into "private property." Then came drones, small, low, and persistent. The FAA's Part 107 regulations established 400 feet as the upper limit for most drone operations, but said almost nothing about the lower limit. This left a question hanging: where does public sky end and private sky begin?

If a drone flies twenty feet above a backyard, is it trespassing or transiting? If it captures imagery from a hundred feet above a fence, is that airspace public or private? No consistent legal answer exists, and that vacuum breeds confrontation between citizens, corporations, and governments alike.

131

Property law depends on boundaries. Air has none. The landmark case United States v. Causby (1946) was the first to challenge the vertical reach of ownership. The Supreme Court ruled that while property owners do not control "the airspace beyond immediate use," they do have rights to the "immediate reaches" above their land. That ruling made sense in an era of low-flying bombers. It makes far less sense in an age of camera-equipped quadcopters. Drones don't just pass over; they hover, observe, and persist, actions the Court never envisioned. Causby gave us a phrase, not a framework. "Immediate reaches" means nothing when technology can occupy them indefinitely. The result is a legal no-man's-land where frustration grows faster than jurisprudence can keep pace.

Drones are the ultimate disruptors of traditional boundaries. They are neither aircraft nor ground vehicles; they operate in a realm that is simultaneously private and public. For homeowners, they represent an intrusion for innovators, an opportunity. For law enforcement, they are tools. For policymakers, puzzles. The FAA regulates for safety, not privacy. State legislatures regulate for privacy, not flight. The courts interpret inconsistently. Local governments, often under pressure from residents, attempt to fill the gaps with ordinances that sometimes conflict with federal law. The result: fragmented authority, competing interests, and ongoing litigation. In short, the nation that conquered the skies has yet to define them.

Corporate delivery systems, aerial mapping companies, and infrastructure inspectors have transformed the low-altitude sky into a lucrative frontier. Additionally, the race to monetize airspace is moving faster than the government's ability to manage it. Amazon, Wing, and other logistics firms are testing drone delivery corridors nationwide, advocating for "aerial highways" managed by automated systems. But who authorizes those routes, the FAA, the state, or the homeowner whose property lies below?

The concept of "common airspace" was meant for open fields and rural routes, not neighborhoods and backyards. When profit soars, privacy exits quietly. Without a clear national framework, the air above every community risks becoming an invisible marketplace, owned by no one, controlled by everyone, and protected by no one.

At its heart, the altitude problem is not just technical; it's philosophical. Who holds sovereignty in a shared sky: the citizen or the state? If the federal government claims all airspace, it effectively limits private recourse against aerial trespass. If private citizens claim control of the space above their homes, innovation is stifled by lawsuits. The solution must strike a balance between safety, sovereignty, and practicality. The law must evolve from territorial ownership to functional use, defining rights not by height but by purpose. For example, a drone conducting a power-line inspection should not be treated the same as one peering into a bedroom window. Intent, not altitude, must define legality.

Every leap in aviation has forced lawmakers to catch up, often too late. When the Wright brothers flew at Kitty Hawk, there were no aviation laws in place. When satellites began orbiting the Earth, no space treaties were in place. The same lag now repeats itself with drones, albeit at a faster pace. Technology accelerates exponentially, legislation crawls. Congressional hearings on drone regulation often sound like time capsules, policymakers struggling to understand innovations already obsolete. By the time a law is drafted, the threat has already evolved. In this sense, the altitude problem isn't just legal, it's existential. Lawmakers are trying to legislate still air in a storm.

Stewardship, equity, and responsibility remain cornerstones of ethical governance, especially in uncharted domains, such as the sky. The principle of care over conquest applies not just to the soil beneath our feet, but to the airspace above our heads. As drones become embedded in commerce, agriculture, disaster response, and national defense, we must resist the temptation to equate accessibility with permissibility. The sky is not a loophole for ethics. Freedom of movement does not absolve the duty of care.

Civic wisdom, drawn from centuries of ethical frameworks, warns that when innovation is pursued without balance, public trust begins to erode. Just as markets require regulation to avoid exploitation, emerging technologies demand public accountability to avoid societal harm. A nation that prioritizes technological ambition over civic responsibility risks losing its moral bearings. The right balance honors both progress and privacy, as well as enterprise and equity, ensuring that our reach never exceeds our principles.

During a policy roundtable, a lawyer joked, "Everyone wants their piece of the sky, but no one wants to pay property tax on it." The laughter was knowing. It captured the absurdity of a legal system trying to tax, regulate, and litigate something that can't be fenced in, bought, or paved over. Yet that absurdity is precisely what happens when laws lag behind life.

To restore order to the air, the nation must establish a comprehensive airspace framework that includes:

- **Defined Altitude Boundaries:** Clarify the distinction between public and private zones below 400 feet.
- **Unified Privacy Standards:** Integrate FAA safety rules with privacy protections under federal law.
- **Local Input Mechanisms:** Allow municipalities to regulate low-altitude activity without conflicting with federal oversight.
- **Rapid Legislative Review:** Mandate regular congressional updates as drone technology continues to evolve.

Without these, the airspace above our heads will remain an unclaimed frontier. And frontiers without law always invite conflict.

In the next section, *"Washington's Whiplash: Patchwork Legislation and Political Inertia,"* we'll examine how political polarization, bureaucratic turnover, and short-term thinking have paralyzed drone policy, leaving innovation unchecked, accountability unclear, and the nation's airspace governed by good intentions instead of good law. Because in the race between lift and legislation, the sky isn't the limit, it's the loophole.

Section 2: Washington's Whiplash: Patchwork Legislation and Political Inertia

"Policy is how a government remembers; politics is how it forgets." Adapted from a National Defense Forum remark, 2022

Every administration promises innovation. Few deliver regulation. Drone policy in America has become a study in whiplash, momentum followed by hesitation, announcements without implementation, and endless hearings that end in handshakes rather than statutes. One year, Congress demands oversight; the next, it demands deregulation. Agencies form committees, release frameworks, then rewrite them after the next midterm election. By the time consensus forms, technology has already advanced two generations. The result is a nation simultaneously overregulated and underprotected. Rules abound, but none align.

The Federal Aviation Administration remains the cornerstone of U.S. airspace management, yet it was never designed for this mission. Its charter was built for pilots, not programmers, for planes, not payloads. The FAA's dual mandate, promoting aviation while ensuring safety, creates constant tension. Encouraging drone innovation means approving more flights; ensuring safety means restricting them. The agency must advance and restrain simultaneously, a bureaucratic paradox that guarantees gridlock.

Part 107, introduced in 2016, was hailed as a breakthrough. It wasn't. While it legalized commercial drone operations, it left critical questions unanswered, including those related to privacy, enforcement, and jurisdiction at low altitudes. The FAA is managing a revolution with the tools of routine. And routine, in the face of exponential change, becomes inertia.

If the FAA is overwhelmed, Congress is divided, not just politically, but also technologically. Lawmakers often lack the technical expertise to legislate emerging systems effectively. The average drone manufacturer updates its firmware faster than Congress updates its policies. Thus, hearings on drone safety regularly veer into debates about unrelated issues, privacy, immigration, and infrastructure funding, while core questions go unanswered. The few members who understand the technology often rotate out of key committees before reforms mature. Each election cycle erases institutional memory, causing progress to end through gradual loss rather than direct resistance.

Drone regulation, like most national issues, has become hostage to polarization. Democrats emphasize privacy and environmental impact. Republicans emphasize innovation and deregulation. Both are partly right and completely stuck. Bipartisan bills stall not because of disagreement over drones, but because of the other issues attached to them. Policy becomes collateral damage in a war of narratives.

Meanwhile, the skies don't wait. Foreign manufacturers flood the market. Municipalities pass their own laws. Enforcement lags behind ambition. The longer Washington debates definitions, the more complex the reality becomes. Drones, unlike voters, don't recognize party lines.

Drone policy is fragmented among various agencies, including the FAA, DHS, DOJ, DoD, and the FCC, each with partial authority and a complex bureaucracy. The FAA governs airspace. The Department of Homeland Security monitors threats. The Department of Defense manages national defense and counter-UAS programs. The Federal Communications Commission regulates radio frequencies. The Department of Justice handles enforcement. In theory, this system ensures checks and balances. In practice, it ensures delay.

Every interagency memo requires coordination. Every coordination requires clearance. By the time approval arrives, the original threat has already moved. The system meant to prevent mistakes now prevents decisions. The result is a bureaucratic Bermuda Triangle, where policy disappears into process and urgency drowns in protocol.

Policy paralysis carries a price tag. Without cohesive drone governance, the United States risks falling behind both allies and adversaries. For example, the European Union has established uniform drone regulations under EASA, streamlining certification and registration across member states. China's centralized model, for better or worse, allows rapid testing and deployment under a single command structure. Israel integrates commercial and defense applications seamlessly, ensuring innovation and security reinforce each other. The United States, meanwhile, remains a patchwork quilt, brilliant in theory, brittle in practice. Political hesitation has transformed airspace into a legislative vacuum. And nature, as well as technology, abhors a vacuum.

Modern governance operates on election cycles, not evolution cycles. Politicians legislate for applause, not adaptation. Drone technology, however, evolves on quarterly timelines, not four-year terms. By the time legislation catches up to the headlines, the headlines have changed. What began as a debate about consumer drones now includes unmanned taxis, delivery systems, and autonomous swarms. Each innovation outpaces the policy vocabulary meant to describe it. Governance cannot afford to move at the speed of comfort.

Powerful private interests further complicate drone policy. Lobbyists representing manufacturers, logistics companies, and data firms shape regulations as much as lawmakers do. Industry influence accelerates approval for commercial operations but slows accountability. Lawmakers eager for innovation often overlook the ethical costs of rapid adoption. The result is deregulation disguised as progress, efficiency without foresight. In Washington, every innovation lobby has a slogan. Few have a conscience. Without ethical checks, policy becomes little more than marketing by another name.

The American public, caught between excitement and unease, has little patience for bureaucratic dithering. Citizens want safety and privacy. Companies wish to have the freedom to innovate. Both blame the government for failing to deliver either. Public trust erodes when policy becomes paralyzed. Each new incident, a near-miss aircraft collision, a smuggling attempt, a foreign surveillance scandal, reignites outrage but never produces reform. Eventually, outrage dulls into resignation. People stop expecting clarity. And that's when the real danger begins. When confusion becomes normal, accountability becomes optional.

Bureaucracies lose their way not from malice, but from fatigue. The antidote is vision, leadership that remembers its purpose even when processes fail. Policy reform will come not from new technology, but from renewed conviction: that governance must keep pace with the governed. Faith in institutions must be earned through action, not slogans. Principle must be paired with persistence. Otherwise, the drone age will be defined not by innovation, but by avoidance.

During a Senate hearing, one staffer whispered to another, "If we form one more drone task force, we'll need air traffic control just to manage the acronyms." The laughter, though brief, was telling. Washington's love of committees has become its coping mechanism. When leaders can't solve a problem, they seek to understand it. At some point, the

Republic must remember that analysis without action is just well-documented avoidance.

To restore control, the nation must align vision and governance through four steps:

- **Permanent Drone Integration Office**: A centralized interagency body with authority, not advisory status, to coordinate policy, enforcement, and innovation.
- **Annual Policy Review Mandate**: Congress must review drone legislation annually to keep pace with technological advancements.
- **Public-Private Ethical Charter**: Industry partners must adhere to standardized privacy and safety codes, enforced through transparent oversight.
- **Technology Literacy Initiative**: Lawmakers must be trained in emerging technologies to bridge comprehension gaps and legislate with clarity.

Good policy requires both urgency and humility, the recognition that waiting for perfect consensus guarantees perpetual delay. The skies are not waiting. Neither should we.

In the next section, *"Lessons from Abroad: EU, Israel, and Japan's Adaptive Frameworks,"* we look outward to nations that have successfully harmonized innovation with regulation. Because sometimes, to overcome Washington's inertia, we must study those who decided that policy should lead the way rather than chase it.

Section 3: Lessons from Abroad: EU, Israel, and Japan's Adaptive Frameworks

"The mark of wise nations is not how fast they invent, but how quickly they adapt.", International Policy Forum, 2023.

Sometimes, the best way to understand America's challenges is to study those who chose differently. While Washington debates definitions, other nations have already codified frameworks, tested responses, and aligned their institutions. The European Union, Israel, and Japan, three vastly different societies, have demonstrated that effective drone governance isn't about size or wealth; it's about discipline. Their models vary, but they share a common principle: policy must evolve in tandem with the machine, not lag behind it.

The European Union took what the United States fears most, centralized regulation, and turned it into an advantage. In 2021, the European Union Aviation Safety Agency (EASA) introduced a uniform set of drone regulations applicable to all member states of the European Union. The system categorizes drones by risk, rather than size or purpose, and applies the same safety and privacy standards across the continent. This framework eliminated the confusion that plagues American policy. A drone operator in France follows the same rules as one in Germany or Spain. While critics call it bureaucratic, Europe's approach has created predictability, the cornerstone of both safety and commerce. Manufacturers know the limits; operators know the expectations. By unifying twenty-seven nations under a single legal authority, Europe did what Washington can't manage across fifty states.

The EU model's genius lies in its simplicity. Instead of regulating by fear, it regulates by function. Drones are grouped into three categories: Open (low risk, minimal regulation), Specific (moderate risk, regulated by operational conditions), and Certified (high risk, equivalent to manned aviation). This stratification allows innovation to proceed where it's safe and imposes control where it's necessary. It's not perfect, no system is, but it works because it strikes a balance between freedom and foresight. The U.S., by contrast, continues to treat all drones as either harmless toys or dangerous weapons, leaving no room for nuance. In policy, extremes don't protect; they paralyze.

If Europe represents unity through regulation, Israel represents agility through necessity. Surrounded by threats, short on land, and high on innovation, Israel has turned its airspace into a laboratory of precision governance. The Israeli Civil Aviation Authority (ICAA) works closely with the Israel Defense Forces (IDF) and private companies. The result is seamless coordination between civilian and military airspace, something the United States continues to struggle with.

Israel's Sky Dome Initiative integrates national radar, counter-UAS systems, and real-time monitoring into a single digital network. The civil and defense sectors share data instantly, enabling an unidentified drone to be tracked, identified, and, if needed, neutralized within seconds. This efficiency comes not from resources, but from mindset, a recognition that bureaucracy is a luxury security cannot afford.

Israel's model thrives because it does not separate innovation from responsibility. Every drone manufacturer must register with both the CAA and the Ministry of Defense. Every operator must undergo basic security training. In return, companies receive expedited testing and deployment approval, accountability exchanged for access. That covenant between industry and government fosters trust, not tension. It's proof that progress doesn't require abandoning caution; it requires earning it.

Japan offers a quieter lesson: stability through structure. Unlike Israel's urgency or Europe's unification, Japan's strength lies in cultural continuity. The Japanese government views technology governance as an extension of social harmony, rather than a means of competition. The Japan Civil Aviation Bureau (JCAB) established one of the world's most methodical certification systems for unmanned aircraft. Every drone flight requires registration, operator identification, and traceability. Although Japan enforces strict regulations, the country maintains one of Asia's fastest-growing drone industries. Transparent and predictable rules encourage companies to invest and innovate.

Despite their differences, Europe, Israel, and Japan share key traits that the United States lacks:

- **Continuity of Vision**: Long-term strategies survive political cycles.
- **Centralized Accountability**: One agency leads; others support.
- **Functional Flexibility**: Regulations scale with risk, not rhetoric.
- **Public Trust**: Citizens understand the balance between safety and privacy.

America's fragmented model, built on overlapping jurisdictions, fails to achieve any of these consistently. It's not that the U.S. lacks talent or technology; rather, it lacks trust in its own institutions to collaborate effectively.

To be clear, America cannot, and should not, copy any foreign model wholesale. Each system reflects unique geography, politics, and culture. But the principles can be translated into unity of purpose, clarity of structure, and accountability in execution. If the United States adopted even half of these traits, its drone policy could evolve from reactionary to resilient. Borrowing wisdom is not weakness; it's maturity.

Throughout history, thriving nations have recognized a timeless truth: broad consultation and cooperative governance are not signs of indecision, but of strength. The American experiment is built on balance, between federal and local, individual and institution, but balance alone is not enough. Without deliberate coordination, balance gives way to gridlock, and principled diversity gives rise to dysfunction.

Humility in leadership is not a weakness; it is the foundation of wisdom. In a complex democracy, listening across sectors, ideologies, and disciplines is not submission; it is stewardship of the public trust. The nations that adapt fastest are not those with the most power, but those with the most cohesion.

During a multilateral drone conference in Brussels, one U.S. delegate joked, "We envy Europe's coordination, but not their coffee." The room laughed, a moment of levity in a sea of legal jargon. The truth, however, was clear. Coordination, not caffeine, keeps a system awake.

If Washington is to close its policy gap, it must internalize the lessons of its peers:

- **Codify Long-Term Policy**: Create legislation that survives leadership changes.
- **Centralize Airspace Authority**: Consolidate oversight under one accountable agency.
- **Adopt Risk-Based Frameworks**: Regulate by potential harm, not hardware.

- **Prioritize Public Education**: Build trust through transparency and training.

- **Institutionalize Adaptation**: Review, revise, and iterate every two years.

Policy should not chase innovation; it should run beside it. The goal is not to mimic foreign systems, but to learn from their clarity, their cohesion, and their courage to act. The United States has the capacity to lead the world in drone governance. What it needs is the will to align its institutions with its ambitions.

In the next section, "Reforming from the Ground Up: Federalism and State Innovation," we bring the lessons home by examining how America's fragmented system can turn its diversity into an advantage by allowing states to pilot reforms the federal government is too slow to test. Because the solution may not come from Washington at all, it may come from the ground beneath the drones.

Section 4: Reforming from the Ground Up: Federalism and State Innovation

"The genius of American government has always been that it lets the states act while Washington debates." " adapted from Justice Brandeis' federalism principles.

America's most significant policy weakness, decentralization, is also its hidden strength. While Washington wrestles with bureaucracy, states retain the freedom to experiment, adapt, and lead. Federalism provides a lasting advantage by enabling local innovation while legislative processes progress slowly elsewhere. In the drone era, where technology evolves faster than federal processes, state and municipal governments have become the testing grounds for national policy. From North Dakota's precision agriculture corridors to Texas' counter-UAS integration zones, states are proving that the air above local communities doesn't have to wait for federal consensus. When the nation hesitates, the states hustle.

Justice Louis Brandeis famously called states the "laboratories of democracy." That phrase has never been more relevant. North Carolina pioneered the use of drones for infrastructure inspection, reducing bridge survey times by 80%. Florida integrated drone mapping into disaster recovery after hurricanes, cutting FEMA processing time nearly in half. North Dakota established one of the first beyond visual line of sight (BVLOS) flight corridors in the nation, an innovation that became a model for future FAA waivers. These initiatives weren't dictated from D.C.; they were developed by governors, engineers, and local leaders who recognized a need before it was noticed. Progress doesn't require permission; it requires initiative.

Every drone flight takes off and lands somewhere local. That reality gives cities and counties both authority and responsibility. Municipalities have enacted ordinances to address noise, safety, and privacy concerns, where federal rules remain unclear. Some regions, such as California and Colorado, have established public-private drone advisory boards to reconcile citizen concerns with commercial interests. These small-scale experiments may lack glamour, but they achieve something Washington rarely does trust. When residents see local leaders regulating with transparency and accountability, compliance follows willingly rather than resentfully. Good governance doesn't need to be grand; it just needs to be grounded.

Federal agencies focus on high-altitude threats and national defense, leaving local authorities to address what's directly overhead: critical infrastructure, sports arenas, public gatherings, and urban corridors. States like Texas, Arizona, and Virginia have stepped in, passing their own counter-UAS legislation to define procedures for drone interdiction within state jurisdiction. While federal law still prohibits most forms of drone interception, these states are shaping the policy conversation, proving that defense can be proactive without overreach. If the federal government defines sovereignty, states define stewardship.

Several states have created "innovation corridors", controlled regions where commercial, research, and public safety drones can operate under shared data and safety frameworks. North Dakota's Grand Sky Complex, Virginia's Mid-Atlantic Aviation Partnership, and Texas's UAS Integration Pilot Program exemplify how structured experimentation can replace guesswork. These corridors facilitate collaboration among academia, industry, and government, eliminating the need for national directives. They represent what America does best: competing locally to improve nationally.

The danger of local control is inconsistency. When every city drafts its own drone law, the result can resemble a jigsaw puzzle missing half its pieces. However, inconsistency can be managed through coordination, rather than control. The National Conference of State Legislatures (NCSL) now provides model policies that help states align on registration, privacy, and enforcement standards while preserving flexibility. This cooperative federalism, where states align voluntarily, offers a template for future national frameworks. Regulation doesn't have to mean restriction; it can mean responsibility.

Stewardship and accountability begin closest to home. The measure of a nation's strength often reveals itself in the competence and integrity of its local leaders, the ones who turn national ideals into daily realities. Leadership at the regional level carries a unique form of trust. These leaders may not control the nation's skies, but they can govern their share of it with foresight, fairness, and a sense of duty to the public good. Protecting one community's privacy, safety, and balance between innovation and restraint may seem small, but collectively, such vigilance defines a nation's character. Ethical governance does not scale upward; it begins locally and radiates outward.

Federalism isn't a flaw in the American design; it is its living strength, a testament that self-government endures when practiced with integrity, humility, and shared responsibility.

The most effective drone regulations include citizen participation. States like Minnesota and Oregon have hosted public drone hearings, where residents could question operators, review flight plans, and learn about safety measures. This grassroots transparency builds both awareness and accountability. When people understand why drones fly, they stop fearing them and start supporting the systems that keep them safe. Public education doesn't weaken security; it strengthens it by turning bystanders into partners.

During a town hall in Oklahoma, a farmer reportedly said, "If Washington can't figure out what's flying over my cornfield, I'll just make my own FAA, the 'Farm Air Authority.'" The room erupted in laughter, but the sentiment was pure federalism. When the center stalls, the edges move. Sometimes democracy needs a little local stubbornness to stay alive.

To institutionalize progress, the federal government should adopt a "ground-up doctrine" that includes four key actions:

- **Incentivize State Innovation**: Offer federal grants for pilot programs and regulatory testing zones.
- **Standardize Safety Benchmarks**: Create a shared national baseline for low-altitude operations.

- **Expand State-Federal Collaboration**: Establish regional coordination offices between the FAA and state aviation departments to enhance coordination and collaboration.
- **Promote Citizen Awareness**: Integrate drone literacy into emergency management and public safety outreach.

Federalism can either fragment or fortify. It depends on whether Washington chooses to coordinate or compete. The drone era demands not only national leadership but also local empowerment. The airspace above every community is not just a technical domain; it's a civic one. And the leaders closest to the ground are often the ones best equipped to guide it.

In the next section, *"The Need for a Drone Doctrine: Setting National Policy Direction,"* we'll bring these threads together, translating lessons from abroad and experiments at home into a unified strategy for the United States. Because without a doctrine to guide it, even the most advanced nation risks drifting: airborne, capable, and directionless.

Section 5: The Need for a Drone Doctrine: Setting National Policy Direction

"Strategy without doctrine is drift; doctrine without purpose is noise." Adapted from *U.S. Army War College Principles of Strategy.*

America has the technology, the talent, and the threat awareness to lead the world in unmanned systems governance. What it lacks is direction.

The United States has doctrines for nuclear deterrence, cyber warfare, and space operations, but none for drones, despite their presence in every domain of conflict and commerce. The absence of a National Drone Doctrine has left the nation vulnerable to inconsistency at home and uncertainty abroad. Without doctrine, strategy becomes reaction. And reaction, in a world measured in milliseconds, is surrendered to delay.

Doctrine is not bureaucracy; it's philosophy in uniform. It defines intent, aligns institutions, and ensures continuity when leadership changes. For drones, doctrine must answer five fundamental questions:

- Who controls the air? (Sovereignty)
- Who safeguards it? (Security)
- Who benefits from it? (Equity)
- Who profits from it? (Economy)
- Who protects the people beneath it? (Accountability)

These are not technical questions; they are moral and strategic ones. Without answers, the nation's approach remains fragmented, reactive, and dangerously slow.

The FAA regulates flight. DHS monitors threats. DoD prepares for conflict. DOJ enforces laws. Yet none of these agencies speaks from a unified script. Each acts as a soloist in a symphony without a conductor. This structural dissonance creates hesitation; every response is delayed

153

by clearance, and every clearance is delayed by caution. A national drone doctrine would do what fragmented policy cannot: define roles, clarify authority, and create command unity between civilian and defense spheres. Command and control is not just a military principle; it's the difference between coordination and chaos.

Doctrine must also include a moral dimension. Unmanned technology blurs ethical boundaries. Distance dulls empathy. Automation accelerates judgment. Autonomy tempts recklessness. A drone doctrine rooted only in efficiency will fail; one grounded in ethics will endure. America's approach must affirm three principles:

- **Human Responsibility**: No algorithm should operate without human accountability.
- **Proportional Use**: Drone employment must strike a balance between necessity and restraint.
- **Transparency**: Citizens deserve clarity on how technology is used to surveil, protect, and govern them.

Stewardship, not control without conscience, but capability with compassion, must guide every operational decision. Faith and freedom both demand it.

A drone doctrine must also establish structural unity. That means creating:

- **National Airspace Integration Council (NAIC)**: A permanent interagency command bridging FAA, DHS, DoD, DOJ, and FCC.
- **Unified Classification System**: Categorize drones by risk, capability, and intent, not arbitrary technical specs.
- **Civil-Military Coordination Framework**: Seamless transition between law enforcement and defense in domestic threats.
- **Technology Review Board**: Independent oversight for AI-enabled or autonomous systems.

Doctrine doesn't expand bureaucracy; it focuses it. It turns overlapping jurisdictions into a single operational rhythm.

As other nations codify their drone strategies, America risks losing its moral and diplomatic leadership. Without a clear doctrine, the U.S. cannot set global norms or defend its ethical standards abroad. A coherent drone doctrine would allow America to lead in international policy, establishing boundaries on autonomous weapon use, surveillance ethics, and export accountability. Global leadership doesn't come from dominance; it comes from discipline.

Doctrines must also connect regulation to innovation. The drone economy is expected to exceed $90 billion globally by 2030, impacting every sector, from logistics to agriculture and defense manufacturing. Without a guiding doctrine, American companies face inconsistent rules, fragmented permits, and limited global influence. A policy that encourages ethical innovation, striking a balance between security and entrepreneurship, can transform governance into a driver of growth. Doctrine, in that sense, is not just a defense strategy; it's an industrial strategy.

Thus, the foundation of any lasting doctrine, whether national or local, must be trust. Trust between citizens and government. Between ambition and accountability. Between progress and principle. Without vision, policy becomes law without life. It may check a box, but it fails to chart a course. And in the vacuum of direction, public confidence erodes.

A clearly articulated national drone doctrine would reestablish that strategic vision, not simply as a technical manual for aerial systems, but as a moral and civic covenant. This framework affirms America's intent not just to shape the sky, but to protect the values anchored beneath it. It would signal to leaders and citizens alike that freedom is not safeguarded by power alone, but by principle, consistently applied, thoughtfully coordinated, and transparently governed.

During a defense policy summit, one retired general quipped, "We've got doctrines for breakfast, lunch, and dinner, but apparently, drones missed the meal." The audience laughed, then grew quiet. Because behind the humor was truth: without a drone doctrine, even the best intentions remain tactical improvisation.

To chart a meaningful path forward, the U.S. Drone Doctrine should rest on five pillars:

- **Unity of Command**: Integrate airspace governance under one executive authority.

- **Moral Accountability**: Embed ethics in every operational protocol.

- **Adaptive Lawmaking**: Require legislative review every 2 years to keep pace with technological change.

- **Public Awareness**: Launch a national education campaign on responsible drone use.

- **International Cooperation**: Lead global dialogue on the governance of unmanned systems.

Doctrine isn't paperwork; it's purpose, codified. It's the difference between drift and direction, between improvisation and integrity.

In the next chapter, *"Corporate Skies: Profit, Power, and Responsibility,"* we shift our focus from government to the private sector, the new air barons shaping the future of flight through innovation, influence, and, too often, indifference. Because in this race to own the sky, power no longer comes from altitude alone. It comes from who controls the data that defines it.

Chapter 7 – Corporate Skies: Profit, Power, and We Responsibility

Section 1: From Garage to Global: The Startup Arms Race

"Every revolution starts in a garage; the challenge is making sure it doesn't end there.", Anonymous venture capitalist, 2018.

The wildfire was spreading fast. Aerial tankers had been grounded for hours. Why? A drone hobbyist, unaware of the fire, had launched a recreational flight two miles out. His drone entered restricted airspace, forcing a halt in aerial suppression. The fire grew. Homes burned. He apologized on Facebook.

The local VFW organized a town hall. Not to scream at him, but to ask: How had we failed to teach civic responsibility? It wasn't about the drone. It was about a missing sense of shared duty.

The story of drones is the story of ingenuity unleashed. What began as hobbyists tinkering in basements and garages has evolved into a multi-billion-dollar global industry, one that blurs the line between play and power. In the early 2010s, as smartphones shrank and sensors became lighter, a wave of entrepreneurs saw the sky not as a frontier, but as a platform.

Startups like DJI, 3D Robotics, and Parrot democratized flight. Anyone with a credit card and a bit of curiosity can now launch what once required a government program. Innovation was no longer limited to institutions. It belonged to individuals. And that democratization changed everything, for better and for worse.

Venture capital poured in faster than policy could catch up. Investors didn't see drones as aircraft; they saw them as data engines. The sky became the next version of the internet, a domain for connectivity, commerce, and control. Startups promised disruption across various sectors, including agriculture, logistics, film, energy, real estate, and security. Their language was bold. Their ethics were ambiguous.

"Move fast and break things" worked fine when applied to code. Applied to the airspace, it became a hazard. The FAA scrambled to define rules, but the pace of funding outpaced the pace of governance. By the time Washington finished one rulemaking cycle, Silicon Valley had already built ten new prototypes. The drone boom wasn't just technological; it was philosophical. It redefined ownership of the sky as the right to innovate, not the privilege to operate.

Every industrial revolution produces new barons. The railroad had Vanderbilt. The steel age had Carnegie. The digital frontier was home to Bezos and Musk. The drone era now has its own titans, executives who think in terms of altitude rather than acreage. Companies like DJI, Skydio, and Zipline turned once-niche engineering feats into global supply chains.

Zipline now delivers medical supplies in Rwanda faster than ambulances can drive. Skydio's AI-powered navigation outperforms human pilots in obstacle avoidance. Each breakthrough redefines what's possible and what's permissible. Corporate ambition has always been America's fuel. The challenge is ensuring it doesn't burn the Republic it powers.

While American startups dream big, Chinese manufacturers build bigger. DJI, now controlling over 70% of the global market, leveraged scale, speed, and state support. U.S. innovators struggled against regulatory hurdles and higher production costs, while China turned its industrial strategy into economic supremacy.

By 2018, American law enforcement and defense agencies were flying drones made by a company based in Shenzhen, ironically, the same country the Pentagon identified as a strategic competitor. That contradiction wasn't malice; it was market logic. When the government stalls, capitalism fills the vacuum, even if it imports risk with the hardware.

Drones don't just fly; they collect. Every image, coordinate, and signal becomes part of a larger data ecosystem. The sky has become a cloud, not of vapor, but of information. Startups realized early that the real money wasn't in selling drones, but in selling what drones see.

Aerial data now fuels industries from precision farming to urban planning. The same camera that films a wedding can also map a power grid, track traffic, or analyze property values. The problem isn't capability, it's custody. Who owns the data gathered from public airspace over private land? Who protects it when it's stored on foreign servers? The law has no altitude chart for information sovereignty.

In the rush to innovate, ethical foresight was treated as a luxury. The early drone culture prized disruption. Breaking norms was celebrated as progress. But not all norms deserve breaking. Safety checks were minimized. Privacy concerns were dismissed as paranoia. By the time the first drone incidents made headlines, near-collisions, airspace violations, and smuggling attempts, the damage to public trust was already done.

The startup creed of "build now, ask later" works until the first lawsuit, or worse, the first casualty. Innovation without introspection becomes arrogance with altitude.

But true innovation demands more than speed or disruption. It demands stewardship. Technology leadership is not measured by how much we can build, but by how wisely we manage what we create. The responsibility to "take care" of our shared environment extends beyond land and water; it applies to the sky and the systems we place within it.

The airspace is not a playground for unchecked ambition. It is a shared trust, one that must be governed with discipline, transparency, and humility. When entrepreneurs and policymakers ignore moral responsibility in pursuit of unregulated progress, technology becomes a shrine to self-interest, a monument to momentum, not meaning.

The pioneers of the drone industry are not villains; they are visionaries. But vision without virtue becomes volatility. What's needed is not less innovation, but more grounded leadership: the kind that weighs long-term impact over short-term gain, and understands that the skies we shape today will define the boundaries of liberty tomorrow.

Drones have transformed lives, from delivering defibrillators in Europe to mapping flood zones in Asia. Yet every noble story coexists with a darker counterpart: unauthorized surveillance, smuggling contraband, violating privacy, or weaponizing consumer hardware. Technology itself is neutral. Application defines morality.

The private sector's challenge is not innovation, it's integrity. If corporations want to claim the skies, they must also claim the burden of protecting those beneath them.

At a 2022 tech summit, a venture capitalist bragged, "We're changing the world with drones!" A retired colonel replied dryly, "Son, I've seen that movie; it ends with me writing the after-action report." The laughter that followed was earned. The room understood: ambition without accountability eventually lands on someone else's desk.

Startups often chase attention. Nations chase endurance. The companies that will define the drone era aren't those who sell hardware; they're the ones who build trust. As the market matures, investors are learning that ethical governance isn't a constraint; it's a competitive advantage.

The next generation of drone leaders will not be defined by how high they fly, but by how responsibly they land.

In the next section, *"Silicon Ambition: When Disruption Replaces Discipline,"* we'll explore how the cultural DNA of Big Tech, the obsession with speed, dominance, and disruption, reshaped drone development, often faster than wisdom could follow because innovation doesn't need more altitude. Sometimes it just requires more accountability.

Section 2: Silicon Ambition: When Disruption Replaces Discipline

"Move fast and break things" sounds bold until what breaks is trust. Industry Commentary, 2019.

Disruption has become Silicon Valley's unofficial religion. The word once meant "to improve the status quo." Now it often means "to destroy it faster than anyone can react." In the race to innovate, caution became a synonym for weakness. Startups were told to "fail fast," but no one explained what to do when failure involved public safety, privacy, or airspace violations.

The tech industry thrives on creative chaos. But the sky is not a sandbox. In airspace, errors have consequences measured not in downloads but in casualties. The problem isn't ambition, it's amnesia. A forgetting that technology serves humanity, not the other way around.

Silicon Valley measures success by velocity, how quickly a product launches, scales, or disrupts the market. But flight, both literal and metaphorical, demands patience. Engineers once spent years certifying aircraft. Drone startups often spend weeks. Testing cycles are compressed. Ethics reviews are optional. Regulatory compliance is treated as an obstacle rather than a requirement.

The faster the climb, the steeper the fall. Innovation without calibration creates turbulence, not progress.

Autonomy has become the holy grail of modern technology. If a drone can think for itself, the argument goes, it can operate anywhere. But autonomy without oversight is not intelligence, it's indifference disguised as innovation. AI-driven drones now make split-second decisions that previously required human judgment, from obstacle avoidance to target tracking.

When algorithms act without understanding, ethics become an afterthought. The same logic that fuels self-driving cars now powers self-flying systems. Yet unlike highways, the sky lacks signage. Without doctrine or discipline, autonomy risks becoming an echo chamber of code. Technology that operates faster than the conscience inevitably collides with consequences.

Today, the modern startup is built on efficiency, minimizing cost, maximizing output, and automating everything in between. Efficiency is good until it becomes an idol. Corporations often use the language of optimization to justify moral shortcuts. When a product fails, they call it a "learning opportunity." When it violates privacy, they call it "data utilization." When it bypasses regulation, they call it "disruption."

Faith and reason both warn against worshiping false gods. In Silicon Valley, the golden calf is efficiency. The pursuit of progress without pause leads to burnout, breakdown, and backlash.

Tech entrepreneurs often claim the government "doesn't understand innovation." They're partly right. But they also misunderstand governance. Regulation is not designed to stop progress; it's meant to prevent collateral damage. Aviation laws were written in blood, every line paid for by a crash, a failure, or a lesson learned too late.

Disruption may build empires. But discipline sustains civilizations. The companies that endure are those that learn to innovate within guardrails, not around them.

Today's technologists often place their confidence not in timeless principles, but in the elegance of code. There's a prevailing belief that data is pure, that numbers do not lie, that algorithms are neutral arbiters of truth, and that automation can outgrow human error.

But data, by its nature, reflects its designers. Algorithms absorb human assumptions with mechanical accuracy. The result? Systems that reproduce bias at scale, cloaked in the illusion of neutrality. This illusion is seductive until a drone makes a false positive or an AI system disproportionately targets a vulnerable population.

Technology does not absolve society of its ethical responsibilities; it amplifies them. When we surrender critical judgment in favor of computational efficiency, we don't evolve, we abdicate. And when ambition goes unchecked by reflection, innovation becomes a monument to hubris rather than a measure of progress.

The pioneers of Silicon Valley often defend themselves with the phrase, "We just build the tools." It's the same defense every blacksmith, chemist, or engineer has used since history began, until the tools were used for harm. The line between invention and complicity is drawn not by design, but by intent. Refusing responsibility for the application is not neutrality; it's negligence.

Innovation carries moral ownership. Once technology leaves the lab, its consequences belong to its creators, whether they claim them or not.

Silicon Valley rarely looks backward. Startups rise and fall so fast that institutional memory resets every quarter. Mistakes are forgotten.

Lessons are archived in venture capital slides rather than being publicly recorded. That's why the same ethical errors recur, new CEOs making old mistakes in sleeker packaging.

Meanwhile, the public forgets just as quickly. Data breaches fade from headlines. Privacy scandals fade from outrage. Consumers trade outrage for convenience faster than regulators can file subpoenas. Innovation thrives when the public forgets. Responsibility thrives only when it remembers.

Discipline is not the enemy of innovation; it is its safeguard. Without boundaries, progress becomes perilous. Without foresight, growth becomes unchecked acceleration. History and ethical governance both remind us that restraint is not a sign of weakness; it is a measure of maturity. It is the wisdom to act not simply because we can, but because we should.

In the age of rapid automation and market disruption, that lesson bears repeating: Better a company with conscience than one that merely conquers markets. Better a policy built on deliberation than one driven by impulse. A society that forgets restraint also forgets responsibility. And when responsibility is removed from the equation, what remains is not innovation, but intrusion.

At a drone expo in Las Vegas, a startup CEO boasted, "Our system can map a city in under an hour." A journalist asked, "Did the city agree to be mapped?" The CEO hesitated, smiled, and said, "We prefer to ask forgiveness, not permission." The audience laughed nervously. Everyone knew how close that philosophy hovers to lawlessness.

The solution is not to slow innovation, but to mature it. Corporate ambition must evolve from disruption to discipline. From breaking rules to building standards. Discipline is not a limitation; it's a long-term strategy.

Every successful aviation era, from the Wright Brothers to SpaceX, paired freedom with frameworks, passion with patience. The companies that endure will be those that build with accountability, not adrenaline.

In the next section, *"Corporate Custodianship: Ethical Manufacturing and Accountability,"* we examine the next evolution of this story, how corporations must grow from innovators into custodians, bearing the moral weight of the systems they create. Because disruption may start a revolution, but only discipline can sustain one.

Section 3: Corporate Custodianship: Ethical Manufacturing and Accountability

"With great power comes great accountability,, Adapted from the timeless truth every leader eventually learns.

Every generation of inventors must decide what kind of legacy it wants to leave: convenience or conscience. The drone industry has reached that inflection point. What began as a race to innovate must now become a mission to steward. Drones are no longer curiosities; they are tools of commerce, defense, and governance. The stakes are too high for neutrality. Corporate success once depended on velocity. Now it depends on virtue. The next era of aerial innovation will belong not to the fastest companies, but to the most responsible.

Ethical manufacturing begins long before the assembly line. It starts with intent, the moral blueprint behind the design. A company that builds for efficiency alone will cut corners where it matters most. A company that builds for humanity will design for safety, privacy, and transparency as non-negotiables. Material sourcing, labor conditions, and environmental impact shape the unseen moral dimension of every aircraft. The goal is not perfection, but integrity: ensuring that the product reflects the principles of the people who built it.

Modern manufacturing is global. And with globalization comes opacity. The average drone is a mosaic of international components, motors from one country, chips from another, cameras from a third. When corporations fail to vet their supply chains, they inherit the ethics of every subcontractor involved. A drone assembled in California may carry sensors from a region accused of forced labor. A firmware update coded overseas may include vulnerabilities unknown to the operator. Accountability cannot stop at the loading dock. Corporate custodianship means tracing responsibility from concept to circuit.

Every drone is a flying computer. Which means every drone is a potential weapon if compromised. Ethical responsibility now includes cybersecurity as much as aerodynamics. Manufacturers must ensure encryption, data protection, and firmware integrity. Yet too many companies prioritize market share over hardening. When the United States banned the use of foreign-manufactured drones in sensitive operations, it wasn't an act of paranoia; it was a lesson in complacency. Because security built in haste is security already breached. A genuine custodian doesn't just ask, "Can it fly?" They ask, "Can it be trusted while it does?"

Quality control ensures performance. Moral control ensures purpose. Corporations must establish internal ethics boards that review not only the functionality but also the implications, asking, "What could this technology do if misused?" The defense industry has long used this practice. Commercial tech firms often avoid it, fearing delay or publicity. Yet ethics isn't bad press; it's good policy. Companies that self-regulate effectively rarely face the burden of external regulation. Transparency earns time, the most valuable currency in innovation.

Consider the dual-use dilemma. A drone designed for agricultural use can also be used to map military installations. A system built for delivery can be modified to deliver explosives. This dilemma haunts nearly every manufacturer. The choice isn't whether to stop innovation, but how to anticipate exploitation. Ethical companies employ "red teams", internal groups tasked with probing vulnerabilities and misuse potential before adversaries do. It's cheaper to fix a flaw in a lab than a scandal in Congress. Maturity in industry means assuming worst-case scenarios and designing for them, not denying they exist.

Work without integrity is effort without honor. No matter the brilliance of the product or the scale of the impact, if the foundation is hollow, the achievement is superficial. This principle is not reserved for elected officials or service members; it applies equally to those in boardrooms and R&D labs. A company that builds technology without conscience is not innovating, it's assembling cleverness without wisdom.

Corporate integrity is not a slogan; it is a standard. It requires humility, the willingness to question assumptions, to elevate long-term responsibility over short-term gain, and to lead not just with ambition, but with accountability. Progress without principle is not advancement, it is regression in disguise. In the realm of public trust and emerging technology, the cost of that deception can be irreversible.

Accountability thrives in light. Corporations that disclose their testing procedures, privacy protocols, and data practices build public confidence. Strategic action, rather than altruism, drives trust, which in turn minimizes the need for lawsuits, protests, and legislation. When people believe a company acts ethically, they are more likely to cooperate with its mission. That's not PR. That's leadership. Transparency is the modern currency of legitimacy.

Some argue that the government should enforce ethics. Others debate that the market will self-correct. The truth lies between. Government oversight ensures standards, but it cannot supervise every circuit. Private corporations, by contrast, have proximity, the ability to embed ethics directly into design. That proximity comes with privilege. And privilege requires principle. Self-governance is the highest form of accountability, and the hardest to fake.

At a manufacturing conference, a CEO was asked how his company ensured ethical compliance. He smiled and said, "We have a three-step system: we design it, we test it, and then we pray." The audience laughed, but only half-heartedly. Everyone knew that prayer without policy is not a plan. Faith may bless good intentions. But accountability requires follow-through.

The companies that will define the next decade of aerial innovation will be those that design with conscience, not convenience. They will measure success not by market capitalization, but by public trust. They will treat every drone not as a product, but as a pledge, a promise that technology can uplift without undermining. Corporate custodianship isn't charity. It's character in motion. And character, once airborne, becomes culture.

In the next section, *"Partnerships That Protect: PPPs in Aerospace Governance,"* we'll explore how the private and public sectors can align, not through competition, but collaboration, to ensure that the skies remain both open and orderly. Because when government brings structure and corporations bring innovation, what emerges isn't bureaucracy or greed. It's resilience.

Section 4: Partnerships That Protect: PPPs in Aerospace Governance

"No mission succeeds alone. The sky is too large for silos.", Military-Aerospace Symposium, 2022.

The drone revolution is too vast, too fast, and too consequential for any single entity to manage. Governments regulate. Corporations innovate. Communities absorb the impact. Without coordination, each operates with partial vision, and partial vision leads to full-scale risk.

Public–Private Partnerships (PPPs) offer the connective tissue between innovation and governance. They unite public authority with private agility, a balance of order and opportunity. Done right, PPPs can turn bureaucratic delay into shared mission execution. Done wrong, they become political theater. The difference lies in discipline: aligning interests without compromising integrity.

Policy sets the boundaries. Practice reveals the gaps. Government agencies are experts at defining mission requirements but often lack the tools, technology, and timelines to meet them. Private industry, by contrast, possesses the speed, capital, and creativity to respond, but not always the guardrails.

The best partnerships emerge when each side acknowledges its weaknesses. Government needs agility. Corporations need accountability. Bringing those strengths together is not a compromise; it's a strategy. The modern aerospace environment demands mission unity, a shared understanding that protecting airspace is both a regulatory and operational duty.

Across the nation, several initiatives are proving that collaboration works when trust is operationalized:

The FAA's BEYOND Program unites state, local, and industry stakeholders to advance drone integration while evaluating safety and privacy implications.

NASA's UTM (Unmanned Traffic Management) initiative brings together academia, software companies, and public agencies to simulate the management of shared airspace.

The Department of Defense's DIU (Defense Innovation Unit) accelerates the transfer of dual-use technologies, bridging Silicon Valley's creativity with the Pentagon's precision and expertise.

Each program shares a critical trait: defined roles with mutual respect. They work not because the government relaxes oversight, but because it modernizes it.

Airspace security is not only a technical problem; it's a trust problem. Without trust between agencies, industry, and citizens, even the best systems fail. Trust is built through transparency, clear expectations, consistent communication, and shared accountability. When each stakeholder knows how and why decisions are made, compliance follows naturally. Public trust, once earned, becomes an invisible infrastructure stronger than any radar network.

Effective partnership is not a sign of institutional weakness; it is wisdom in action. True collaboration recognizes that no single sector holds all the answers. Government without private innovation risks rigidity. Private industry without public guardrails risks recklessness. Together, they can strike a balance between progress and protection, ambition and accountability.

Sound public-private partnerships must be guided by more than quarterly reports or election cycles. They should serve the common good, not corporate convenience or political credit. At the heart of principled collaboration is a shared mission: to protect the public, preserve democratic institutions, and promote responsible innovation. When both sectors align on that vision, partnership becomes more than functional; it becomes foundational.

Not all partnerships succeed. Some collapse under bureaucracy. Others dissolve under greed. The common failures stem from three causes:

- **Ambiguous Authority**: When no one knows who's in charge, accountability disappears.

- **Unequal Benefits**: When one side profits and the other justifies, trust erodes.
- **Reactive Postures**: When collaboration happens only after a crisis, learning is expensive.

The antidote is clear mission design: define goals, measure outcomes, and share both the credit and the consequences. Partnerships, like aircraft, require maintenance. Without attention, even the best ones crash.

Universities and research institutions play a crucial but often overlooked role in aerospace governance. They serve as neutral ground, where science meets policy, and innovation meets ethics. Programs at MIT, Embry-Riddle Aeronautical, and Texas A&M are already developing interdisciplinary frameworks that combine law, engineering, and public administration. These institutions act as the conscience of collaboration, ensuring that discovery doesn't outpace discernment. Knowledge shared is risk reduced.

Beyond safety and security, PPPs also drive economic resilience. Collaborations in drone logistics, counter-UAS systems, and infrastructure monitoring have generated thousands of jobs nationwide. Every partnership becomes an investment in both prosperity and preparedness. A nation that coordinates its innovators strengthens its workforce as much as its defense. Public policy should not merely permit collaboration; it should reward it.

For PPPs to succeed, the public must be more than an observer; it must be a participant. Awareness campaigns, community briefings, and transparent reporting help citizens understand how drones are used, monitored, and regulated. The public's right to know must coexist with the state's duty to protect. When both are respected, fear subsides and cooperation grows. Civic engagement transforms passive concern into active partnership.

During a joint workshop, a federal official said, "We need industry to think like the government." An engineer shot back, "If we do that, nothing will launch." The room burst into laughter, but the point was clear. Partnerships don't require uniformity. They require unity. Different mindsets, when aligned to the mission, become a force multiplier.

The future of aerospace governance depends on institutionalizing partnership as a principle, not an afterthought. To achieve this, the nation must:

- **Formalize PPP Frameworks**: Establish standardized templates for risk-sharing and data governance.
- **Expand Joint Training**: Integrate public and private operators into unified simulation and certification programs to enhance interoperability and collaboration.
- **Embed Accountability Metrics**: Measure outcomes not just in efficiency, but in equity, ethics, and resilience.

- **Ensure Continuity Across Administrations**: Protect partnerships from political turnover through statutory stability.

Partnership is not a temporary solution. It is the architecture of sustainable governance.

In the next section, *"When Innovation Meets Obligation: Redefining Profit with Principle,"* we conclude this chapter by examining how corporations can leverage ethics as an advantage, proving that long-term success isn't built solely on disruption, but on discipline and duty. Because the future of flight will not be written by those who fly highest, but by those who land responsibly.

Section 5: When Innovation Meets Obligation: Redefining Profit with Principle

"Profit without principle is power without purpose." Adapted from Mahatma Gandhi's Seven Social Sins.

Every revolution eventually faces its reckoning. The drone industry, once celebrated as the purest expression of ingenuity, now stands at that threshold. Innovation has delivered breathtaking capability, yet it has also introduced ethical turbulence. The question is no longer "Can we fly?" but "Should we, and how?" Profit fueled the ascent; now purpose must guide the flight. The next frontier of leadership will not be defined by who dominates the market, but by who earns the public's trust.

For two decades, the tech sector operated under a single doctrine: growth at all costs. That creed worked when the product was digital. It falters when the product is physical and airborne. A drone isn't an app. It can crash, collide, or surveil without consent. The consequences of innovation now extend from cyberspace to civic space. True capitalism, like true democracy, evolves with accountability. Profit without responsibility is extraction. Profit with responsibility is stewardship.

And so, the challenge is to redefine success not as domination, but as durability, the ability to innovate without eroding the social contract that sustains freedom.

In an age of transparency, ethics is not a cost; it's a currency. Consumers, investors, and governments are increasingly rewarding companies that demonstrate integrity, sustainability, and security in design. Drone manufacturers that embrace ethical standards, from transparent data policies to sustainable sourcing, attract not just customers, but confidence. Markets, like airspace, favor clarity. When corporations communicate their moral compass, the public invests not just in their products, but in their principles. Profit earned with integrity compounds. Profit earned without it corrodes.

History is unkind to those who confuse early success with moral immunity. From railroads to social media, industries that outgrew their ethics eventually faced collapse, not from external enemies, but from internal arrogance. For the drone sector, complacency is the new risk of crashes. The temptation to prioritize margins over meaning often yields short-term victories but creates long-term vulnerabilities. If innovation outruns obligation for too long, the public will respond, through regulation, boycotts, or indifference. And nothing grounds an industry faster than lost trust.

Enterprise and ethics are not opposing forces; they are complementary foundations of a stable society. A corporation led by integrity can thrive with resilience and public trust. One driven solely by unchecked appetite will eventually undermine itself, consumed not by competition but by its own contradictions.

Moral obligation in business isn't about ideology; it's about execution. It's about treating data as a civic trust, labor as human capital with dignity, and the environment as a long-term asset, not a disposable resource. The conscientious entrepreneur understands that purpose and profit must coexist. One without the other invites imbalance, both socially, environmentally, and economically. The future of enterprise will belong not just to the bold, but to the principled.

In the 20th century, Return on Investment was measured in dollars. In the 21st century, it must also be measured in Return on Integrity. Every company should track three forms of profit:

- **Economic Profit** – Did we earn revenue?
- **Social Profit** – Did our work strengthen society?

- **Moral Profit** – Did we act with conscience and fairness?

When these metrics align, capitalism becomes constructive rather than corrosive. Leaders who internalize this model will not just survive in the drone age; they will lead it.

Leadership in aerospace and technology now demands a dual mindset: one that is both visionary and virtuous. The ability to dream must be matched by the discipline to discern. Principle-driven leaders ask hard questions:

- Should we collect this data?
- Should we automate this decision?
- Should we deploy this system in a way that changes how humans live or work?

These are not anti-innovation questions. They are pro-human ones. And answering them defines whether progress uplifts or undermines. The future of flight will not be written by those who invent fastest, but by those who lead with integrity longest.

Compliance is the floor of corporate ethics. Character is the ceiling. A company that obeys the law is a functional entity. A company that honors principles is foundational. When corporations view ethics not as a matter of regulation but as a matter of reputation, their operations transform from reactive to resilient. Integrity becomes a strategic advantage, not a checkbox. In aviation, a strong moral compass is as essential as a stable gyro. Lose either, and everything spins.

During a venture capital roundtable, an investor asked, "What's your company's exit strategy?" The CEO replied, "We don't have one; our conscience won't let us bail before we fix what we broke." The room chuckled, but the message landed. Ethical leadership is not about perfect behavior. It's about refusing to leave chaos behind for someone else to clean up. That's not a weakness. That's maturity.

Corporations shape culture as much as governments shape laws. Every drone that flies, every dataset stored, every sensor deployed is a reflection of the values embedded in its creation. Building a culture of conscience means embedding moral reflection into every process, design, procurement, testing, and deployment. It means asking not only "What can this do?" but "What should this never do?"

That culture cannot be outsourced. It must be cultivated by leaders who believe that innovation without empathy is simply engineering in search of a soul.

The companies that survive the turbulence ahead will be those that rediscover the moral clarity that built American enterprise in the first place: responsibility before reward, service before status, purpose before power. When innovation meets obligation, capitalism finds its conscience again. And when it does, the skies, both literal and symbolic, will belong not just to those who fly highest, but to those who fly honorably.

Innovation made the sky accessible. Ambition made it profitable. However, opportunism has now made it dangerous. For every breakthrough in aerial technology, a darker counterpart emerges, a mirror threat that adapts faster than the law, costs less than defense, and is quieter than detection.

In the next chapter, "Digital Militias: Crime, Cartels, and the New Asymmetric Threat," we move from creation to confrontation, exploring how non-state actors, criminal syndicates, and rogue technologists have weaponized the very systems once hailed as tools of progress. Because in the age of drones, every altitude advantage can be turned on its head. For every aircraft designed to deliver hope, there's another learning how to deliver harm.

Chapter 8 – Digital Militias: Crime, Cartels, and the New Asymmetric Threat

Section 1: The Cartel Cloud: Drones as Smugglers and Scouts

"Technology doesn't choose sides, people do.", Homeland Security Briefing, 2022.

Seventeen-year-old Marcus didn't mean to hack the city's drone registry portal. He was just exploring. But the loophole was there, plain as day, an unencrypted admin password hidden in the page source. With a few keystrokes, he could've grounded every municipal drone in his county.

He reported it instead.

City Hall gave him a certificate. His principal gave him detention for "unauthorized activity." He sat there wondering: *If the people building these systems don't understand the tech, who's actually in charge?*

The same accessibility that empowered entrepreneurs has now emboldened criminals. Consumer drones, once symbols of creativity, have become silent partners in smuggling, reconnaissance, and organized crime. From the deserts of Arizona to the jungles of Colombia, the sky has become contested terrain. And the new players aren't nations, they're networks: agile, anonymous, and armed with off-the-shelf innovation.

This is the cartel cloud, an invisible fleet of modified consumer drones used to move drugs, money, and surveillance data across borders faster than any patrol or radar can react.

The U.S. Border Patrol reported more than 10,000 drone incursions along the southern border in 2023, a 900% increase from just five years earlier. Most of these flights weren't sophisticated military systems; they were consumer-grade UAVs repurposed for criminal enterprise. Cartels learned quickly that drones could do what mules and lookouts could not:

- Fly undetected over surveillance zones.
- Monitor patrol movements in real time.
- Deliver payloads with precision to pre-mapped coordinates.

A $1,200 drone can carry up to $40,000 worth of narcotics. A swarm of five can outmaneuver ground units entirely. When air superiority costs less than a used pickup truck, the balance of power shifts.

The first wave of cartel drones was crude, modified hobby systems with taped-on containers and short battery lives. But adaptation came fast.

Now, drones operate as aerial scouts, mapping the movement of law enforcement convoys and alerting traffickers via encrypted apps. They hover over checkpoints to track agent rotations, identify weak patrol schedules, and even record infrared footage at night.

Cartel logistics networks have transformed the sky into an operational dashboard, utilizing low-cost UAVs to coordinate convoys, schedule drop-offs, and evade surveillance. The drone isn't just a tool. It's an ecosystem.

Cartels coordinate drone flights with remarkable precision. Operators launch from Mexican territory while handlers on the U.S. side recover the payloads, often drugs, cash, or encrypted drives containing trafficking data. The entire process takes minutes. No vehicles. No border crossings. No human footprints.

This method has not only reduced cartel risk, but it has also multiplied their efficiency. Traditional interdiction methods, cameras, fences, and sensors, can't stop what flies above them. When geography loses meaning, sovereignty loses traction.

What began as smuggling soon evolved into an offense. Reports from Mexican authorities document drones carrying fragmentation grenades and improvised explosives, used in territorial battles between rival cartels. The Jalisco New Generation Cartel (CJNG) in particular has demonstrated tactical proficiency, coordinating multi-drone attacks during raids and ambushes.

Each drone is cheap, disposable, and deniable. For cartels, drones have become the poor man's air force, asymmetric weapons with strategic precision and political ambiguity. When criminal organizations acquire control of the air domain, the battlefield expands without formal declaration.

Traditional warfare relies on identification, flags, insignia, and transparent command chains. Criminal drone operations thrive on anonymity. When a drone crosses the border, there's no manifest, no radio signature, no pilot to prosecute. Tracing the origin becomes a technical labyrinth, often ending in ambiguity.

Cartels exploit this legal vacuum. If a drone crashes, it loses hardware, not humans. If it's detected, they switch to a different frequency or brand. Law enforcement can't retaliate against a faceless fleet.

Agents on the border describe it as fighting ghosts. Drones appear on radar for seconds, vanish behind ridges, and reappear miles away. By the time interception teams mobilize, the payload has been delivered or the drone has been destroyed remotely.

Counter-UAS tools, such as jammers and net guns, are subject to jurisdictional limitations. Many agencies can't legally deploy them due to FAA restrictions on interference with navigable airspace. The result is operational paralysis: the ability to recognize the threat but not take action to stop it. It's not that law enforcement lacks courage. It lacks authority.

The defense of a nation demands more than tactical superiority; it requires moral clarity. Governments have the authority to protect, but that power must be exercised with precision, not excess. The mission is not to perfect control, but to preserve liberty while ensuring safety.

The fight against the cartel cloud is not only strategic, but also ethical. How do we safeguard without suffocating civil freedoms? How do we secure our borders without militarizing the skies above our communities?

The challenge of today involves maintaining constant vigilance while upholding ethical standards. Strength without accountability breeds fear, not freedom. The goal is not dominance of the air, it is the restoration of accountability beneath it. A sky protected is not a sky surveilled, but a sky stewarded for the public good.

During a border briefing, a weary agent quipped, "We've gone from chasing coyotes to chasing crows with batteries." The laughter masked exhaustion. Every new technology demands new tactics, new budgets, and new burdens. Yet the mission remains the same: protect the line, uphold the law, and go home alive. Sometimes humor is the only thing left when policy can't keep pace with progress.

To confront the cartel cloud, the United States and its partners must build an integrated air-defense posture for domestic security, one that includes:

- **Unified Air Surveillance**: Shared radar and data systems between CBP, DHS, and DoD.

- **Expanded Counter-UAS Authority**: Legal updates allowing active drone interdiction.

- **Public-Private Partnerships**: Collaborations with telecom and software firms for early detection algorithms.

- **International Intelligence Sharing**: Binational task forces for rapid attribution and interdiction.

The sky has become the new border. And borders, like freedom, demand constant defense.

In the next section, *"The Hacker's Hangar: Cyber Manipulation of UAV Networks,"* we descend from physical smuggling to digital infiltration, where hackers, mercenaries, and rogue coders exploit the software seams of unmanned systems. Because the next great air war won't begin with missiles, it'll start with malware.

Section 2: The Hacker's Hangar: Cyber Manipulation of UAV Networks

"He who controls the signal controls the sky.", Cyber Defense Symposium, 2021.

The new drone wars aren't fought with firepower alone. They're fought in the shadows of the spectrum, the frequencies, firmware, and digital backdoors that determine who commands the craft. In this domain, the hacker replaces the pilot, and the keyboard becomes a weapon.

Every drone, from a hobbyist quadcopter to a military-grade system, operates on code. And wherever there is code, there is vulnerability. The hacker's hangar is not a warehouse; it's a Wi-Fi signal, a proxy server, a laptop with the right exploit. From basements in Eastern Europe to cyber cafés in South America, drone hijacking has become the latest frontier in asymmetric warfare.

Modern drones rely on GPS, radio frequencies, and proprietary control links. But these systems were built for convenience, not conflict. Hackers exploit this mismatch. They jam signals, spoof coordinates, and hijack control channels, turning precision tools into unpredictable threats.

In 2024, cybersecurity researchers demonstrated that an off-the-shelf laptop could commandeer a commercial UAV within 200 meters by mimicking its GPS handshake. Cartels and cyber-mercenaries took notes. The same tools once used to test security are now traded in encrypted forums, packaged with user guides, and sold for cryptocurrency. A $900 drone can be weaponized by code costing less than $9.

Drone hijacking follows a disturbingly simple pattern:

- **Interception** – Identify the control signal and frequency.
- **Spoofing** – Clone the handshake protocol to simulate the legitimate controller.

- **Override** – Insert a stronger signal or packet sequence to seize command.

- **Payload Manipulation** – Redirect the drone or extract its data.

These steps take minutes. Once control is gained, a hacker can crash the drone, reroute its payload, or turn its camera into a live surveillance feed, all while the real operator watches helplessly. In cyberspace, visibility doesn't equal control; it only gives the illusion of it.

What began as digital mischief has evolved into a mercenary market. Freelance hackers now rent their expertise to cartels, insurgent groups, and private intelligence firms. These actors don't need fleets; they need access. A single compromised drone in a government fleet can expose entire networks, GPS coordinates, or operational patterns.

The rise of cyber-UAS mercenaries blurs every line between warfare and crime, between combatant and contractor, between attack and espionage. In one documented case, a municipal police drone monitoring a protest was digitally hijacked mid-flight and redirected to crash into its own command vehicle. It wasn't sabotage for profit. It was a spectacle for ideology.

The most sophisticated cyberattacks don't jam or hijack; they lie dormant. Adversaries infiltrate drone firmware at the manufacturing or supply-chain level, embedding "sleeping" code that can activate months later. This supply-chain infiltration is particularly dangerous because it hides inside legitimate updates and vendor maintenance cycles.

Operators never suspect compromise until systems start behaving irregularly, freezing mid-air, leaking data, or relaying false telemetry. It's the perfect crime: invisible, deniable, and profitable. A foreign intelligence service doesn't need to smuggle weapons across borders. It only needs to ship components.

Many manufacturers turn a blind eye to firmware vulnerabilities, preferring speed to security. Open-source APIs and cloud connectivity enable integration, but also pose a risk of intrusion. When corporations treat cybersecurity as a software patch rather than a core principle, they hand adversaries the keys to the kingdom.

And unlike traditional weapon systems, commercial drones lack classified encryption. Their vulnerabilities are published in forums and GitHub repositories. Neglect, not malice, has become the hacker's greatest ally.

Even when a drone is identified as having been hacked, tracing the attacker can be a nightmare. Most exploits bypass anonymized VPNs, dark-web command nodes, or hijacked cloud servers across multiple jurisdictions. Digital forensics can determine how the attack occurred, but rarely who carried it out.

That's why prosecutions for drone cybercrimes remain in the single digits worldwide. Meanwhile, law enforcement agencies operate under legal constraints that prevent them from engaging in proactive cyber defense. The paradox is stark: the hacker can attack from anywhere, but the defender can only respond within borders. In cyberspace, sovereignty becomes a suggestion.

Vigilance begins with stewardship, and in the digital age, that stewardship extends to every line of code. Cybersecurity is not just a technical obligation. It is a covenant of trust between developers and users, as well as between institutions and the public. Digital infrastructure underpins modern governance and commerce. To safeguard it is to protect more than systems; it is to defend transparency, stability, and democracy itself.

Responsibility in the digital realm is not optional; it is foundational. Because when connectivity outpaces conscience, chaos rushes to fill the void.

At a cybersecurity conference, an analyst joked, "You know you're losing when the drone you're defending uploads its own ransom note." The room erupted, half in laughter, half in recognition. The threat is absurd only until it's real. Then it becomes Tuesday.

Mitigating cyber-UAS threats requires more than patches and policies. It demands a national posture of digital resilience:

- **Zero-Trust Architecture** – Assume Compromise and Design for Containment.

- **Real-Time Threat Sharing** – Establish a unified drone cybersecurity operations center linking federal, state, and industry data.
- **Firmware Vetting and Certification** – Mandate third-party audits before market release.
- **AI-Driven Intrusion Detection** – Use machine learning to identify abnormal flight behavior.
- **Cyber Defense Training** – Develop a dedicated cadre of "digital pilots" trained in both aviation and cybersecurity.

The next frontier of homeland defense isn't just physical control of the airspace. It's the cognitive control of the code that commands it.

In the next section, *"Law Enforcement at a Disadvantage: Legal and Technical Constraints,"* we examine how outdated statutes, fragmented authority, and bureaucratic hesitation have left agencies reactive rather than proactive, outpaced by criminals and outmaneuvered by code. Because in this war, the gap isn't just technological. It's institutional.

Section 3: Law Enforcement at a Disadvantage: Legal and Technical Constraints

"You can't win a 21st-century fight with 20th-century rules.", DHS Field Officer, Southwest Sector, 2023

The United States has one of the most sophisticated airspace systems in the world, but also one of the most legally restricted when it comes to defending it. For decades, air regulation focused on safety rather than security. The Federal Aviation Administration (FAA) governs the sky as if it were a neutral zone, designed for peaceful use, commercial expansion, and freedom of navigation.

That framework made sense when only trained pilots were allowed to fly. However, the democratization of flight, now accessible to anyone with $500 and a smartphone, changed the equation overnight. The law, however, stayed grounded.

The result is a dangerous paradox: federal agencies can track a drone in real-time, but in most cases, they cannot legally engage, jam, or deactivate it without violating the very laws they're sworn to uphold.

Title 18 of the U.S. Code classifies any interference with aircraft as a federal crime, and the law defines drones as aircraft. Unless a specific agency receives authorization under Title 6 (Homeland Security) or Title 10 (Defense), no legal basis exists for engaging in counter-drone activities.

Local and state police, therefore, face an impossible dilemma. They can see the threat. They can document it. They can even identify the operator. But they can't stop it.

One Border Patrol agent summarized it bluntly: "We're watching the sky burn while waiting for permission to put out the fire."

The American homeland operates under a maze of overlapping jurisdictions. The FAA manages airspace. The FCC manages radio frequencies. The DHS coordinates domestic security. The DoD oversees national defense. And the DOJ handles prosecution, when and if the evidence meets statutory thresholds.

Each has a piece of the puzzle. But no one owns the total picture.

Criminals, meanwhile, exploit these seams effortlessly. They operate in the gray zones, below 400 feet, beyond line of sight, between regulatory boundaries, where no single agency has both the authority and capability to act. In this bureaucratic fog, drones not only evade detection but also thrive. They evade accountability.

Technology moves in months. Legislation moves in years. By the time Congress debates new drone provisions, the market has already shifted to the next generation of hardware and software. Even when laws are passed, implementation often lags due to administrative red tape, budgetary constraints, and litigation.

The Preventing Emerging Threats Act (2018) granted limited counter-UAS authority to certain agencies but left massive gaps at the state and local levels. In effect, America created a two-tier defense:

- Authorized watchers who can act.
- Everyone else who can only observe.

In asymmetric warfare, observation without intervention is surrender.

Even when the equipment is available, training often isn't. Counter-drone systems require technical expertise, rapid coordination, and disciplined escalation protocols. Most local agencies lack the resources or personnel to maintain these skills.

A radar display is of little value if no one can interpret it. A jammer is useless if the operator doesn't know the difference between a rogue drone and a civilian delivery craft.

The Department of Homeland Security has made strides through the C-UAS Working Group. But scalability remains the issue. There are over 18,000 law enforcement agencies in the U.S., most of which have zero counter-UAS training. The gap isn't ignorance. It's infrastructure.

Every new authority granted to law enforcement invites public scrutiny, and often, political backlash. Civil liberties groups raise valid concerns about surveillance, privacy, and potential misuse. But this debate has paralyzed decision-making at precisely the moment when clarity is most needed.

Leaders hesitate, fearing optics more than outcomes. The result is a patchwork of pilot programs, waivers, and temporary authorizations that expire before they can be fully evaluated and assessed. A security policy cannot thrive in a state of permanent probation.

Power must always be balanced by principle. Granting counter-drone authorities is not a blank check for control; it is a charge of responsibility. It entrusts leaders to act only when necessary, to protect without overreach, and to preserve both safety and freedom.

The absence of authority does not define good governance; rather, it is the ethical use of authority that matters. Power applied with integrity safeguards liberty. Power applied without restraint erodes it.

The legitimacy of any system depends on trust, trust in the institutions, in the individuals trained to lead, and in the values that guide their decisions. When leaders equip their people with both skills and ethical clarity, authority becomes a protection, not an oppression.

Other nations have not been so hesitant. The United Kingdom, France, and Israel have integrated counter-UAS policies that empower local forces to neutralize rogue drones within defined parameters. They treat the air as both a commons and a command, shared, but safeguarded.

By contrast, the U.S. relies on centralization and caution, favoring litigation over delegation. That model may prevent misuse. But it also prevents readiness. Freedom without structure is fragility disguised as virtue.

Law enforcement cannot be everywhere. And radar cannot see everything. The public must become a force multiplier, eyes and ears attuned to suspicious drone behavior. Programs like "If You See Something, Say Something" should expand to include aerial awareness.

Communities near airports, power plants, and ports can become early-warning nodes in the national detection grid. The front line of defense has always been local. The sky doesn't change that truth.

During a training seminar, an instructor told officers, "Remember, you can't shoot it down, not even if it's spying on your barbecue." The class laughed, half in disbelief, half in frustration. It was a joke, but not a lie.

In most jurisdictions, destroying a rogue drone could cost an officer their badge faster than it costs a suspect their drone. In that irony lies the entire policy failure.

The solution lies in incremental authority expansion and shared responsibility:

- **Modernize Title 18** to distinguish malicious UAV interference from legitimate law enforcement defense.
- **Expand Counter-UAS Permissions** to vetted state and local agencies through certification and oversight.
- **Standardize Use-of-Force Protocols** for aerial interdiction.
- **Invest in Training Infrastructure** at the national level, funded, certified, and sustained.
- **Integrate Public Reporting Networks** with real-time incident databases.

These measures don't militarize the sky. They modernize accountability beneath it.

199

In the next section, *"Lessons from the Frontline: Counter-UAS Field Cases,"* we turn to real-world examples, from urban stadiums to international borders, where courage, coordination, and creativity have filled the policy void. Because in every successful defense, the common denominator isn't technology or funding. It's leadership.

Section 4: Lessons from the Frontline: Counter-UAS Field Cases

"Every drone encounter is a leadership test disguised as a technical problem.",
Joint Interagency Task Force briefing, 2024.

No event tests American homeland security coordination like the Super Bowl. For one week each year, federal, state, and local agencies converge into a single operational organism, a living example of what counter-UAS cooperation looks like when done right.

In 2023, the Department of Homeland Security and NORAD established a Temporary Flight Restriction (TFR) zone over Glendale, Arizona. More than 30 federal, state, and local entities were integrated into a joint operations center, featuring real-time air tracking, RF detection, and rapid-response teams. Despite more than 70 unauthorized drone attempts, none breached the inner security ring.

The secret wasn't technology alone; it was discipline, unity of command, and apparent legal authority delegated in advance. When agencies train together before a crisis, they don't have to negotiate during one.

In Eastern Europe, where NATO's eastern flank meets constant probing from Russian drones, Polish and Lithuanian forces have become masters of rapid adaptation. When small quadcopters began crossing the border to harass logistics routes in 2022, Poland responded with a layered defense doctrine, a fusion of sensor grids, local command autonomy, and civilian reporting apps.

Villagers living near the front line were given simple mobile tools to report drone sightings directly to regional command centers. The result: intercepts rose by 40% over six months, and the average response time fell from 10 minutes to 3. Poland's lesson is timeless: national security is strongest when it is locally rooted.

Along the U.S.–Mexico border, customs and military teams have turned parts of South Texas into a real-world counter-UAS laboratory. At the Joint Interagency Task Force South Test Range, operators simulate cartel drone patterns and test interdiction protocols with live hardware.

Here, CBP agents train alongside Air Force and National Guard units to identify, track, and neutralize rogue aircraft within seconds. When a modified DJI drone breached the range in 2024, the team neutralized it through RF takeover, not destruction, preserving forensic evidence for future analysis. That single decision led to the arrest of a cross-border operator in Laredo within 48 hours.

Lesson learned: every drone is both a threat and a data point. Destroy too quickly, and you erase the trail.

In late 2022, a series of small drone intrusions near power substations in North Carolina and Washington state raised a frightening possibility: that critical infrastructure had become a new testing ground for sabotage. The drones were never armed, but their flight paths revealed intelligence-gathering patterns, including photographing transformers, measuring distances, and testing response times.

Investigators later found that the operators had downloaded open-source mapping software and shared data through encrypted channels linked to foreign actors. The takeaway: in the age of open data, hostile actors don't need sophisticated satellites, just public Wi-Fi and patience. Infrastructure security must now treat curiosity as a potential precursor to conflict.

In every successful incident response, a pattern emerges: the presence of a leader who can decide without waiting for consensus. At the Super Bowl, that leader was the Joint Operations Commander who ordered a rapid shutdown of local RF bands for five minutes to protect a critical handoff, a bold call that required risk tolerance and trust.

At the Polish border, it was a Lieutenant who ignored protocol to act on a farmer's cell phone report and downed a drone carrying live explosives. In Texas, it was a CBP sergeant who chose to capture rather than destroy. Different names, same trait: courage within constraints.

Most lessons from the field share one uncomfortable truth: our doctrine is still catching up to our data. Reports are often classified, siloed, or inconsistent between agencies. Without a unified after-action database, each incident becomes a stand-alone lesson that dies on someone's hard drive.

Doctrine requires memory. Without shared institutional memory, we repeat mistakes in new zip codes. Building a national counter-UAS lessons-learned repository, accessible to both federal and local partners, would turn field chaos into strategic clarity.

Faith teaches that wisdom is refined in fire. James 1:2–4 urges us to "consider it pure joy when you face trials of many kinds, because the testing of your faith produces perseverance." For front-line defenders, that perseverance isn't abstract; it's daily discipline.

They stand between uncertainty and order, armed with training that's often outdated and authority that's always contested. Their faith is not only spiritual, it's operational faith in each other, in their mission, and in the idea that the law they serve will eventually catch up to the threat they face.

During a joint exercise, a soldier yelled, "Sir, we've got multiple bogeys on radar!" The commander deadpanned, "Relax, it's Amazon Prime Day." Laughter rolled through the command post because in a world where commerce and combat share airspace, humor is how you breathe through ambiguity.

Five field lessons stand out:

- **Preparation beats technology**. You can't deploy what you don't train.

- **Local intelligence is strategic intelligence**. Citizens see first what sensors see last.

- **Leadership must own ambiguity**. Waiting for certainty is a luxury the enemy won't grant.

- **Evidence has value**. Neutralize smartly; capture when possible.

- **Doctrine is discipline on paper**. Without it, lessons die with rotations.

These truths apply whether the sky is over Phoenix, Poland, or Pyeongtaek.

In the next section, *"Controlling the Uncontrollable: Next-Generation Interdiction Tactics,"* we look ahead to the future fight, where AI autonomy, directed-energy defenses, and global coordination will determine whether order prevails over chaos. Because the sky is not lost, it's simply waiting for leaders bold enough to reclaim it.

Section 5: Controlling the Uncontrollable: Next-Generation Interdiction Tactics

"The goal isn't to control everything in the sky, it's to ensure that nothing controls us.", NORAD *Strategy Brief, 2024.*

Every commander learns this early: control is an illusion sustained by discipline. The modern sky, crowded with drones, sensors, and algorithms, has erased any hope of total mastery. Yet that doesn't mean surrender. It means adaptation.

The mission now is to build systems that manage uncertainty, rather than pretending to eliminate it. In the age of digital militias and AI-assisted warfare, interdiction must evolve from a reactive to a predictive approach. Faster missiles will not define the future of air defense; smarter decisions will.

Artificial intelligence has already entered the counter-UAS arena. Machine-learning platforms now analyze radar feeds, acoustic signatures, and telemetry data in real-time, predicting hostile intent before human eyes can detect it. The technology doesn't just respond. It anticipates.

In field tests across the Southwest, predictive defense systems correctly identified 92 percent of rogue drones before they entered the airspace. But this success introduces a paradox: the more we automate defense, the less human judgment remains in the loop. Victory comes with a question: who decides to pull the digital trigger?

That question foreshadows Chapter 9.

Kinetic interception, guns, nets, and missiles work, but it's expensive and imprecise. The next generation belongs to directed-energy systems, which include high-powered microwaves and lasers that can disable drones mid-flight without causing collateral damage.

The U.S. Air Force's THOR (Tactical High-Power Operational Responder) and the Army's DE M-SHORAD units represent this shift. They offer silent precision and reusable defense. Still, they're not magic. Directed-energy weapons require immense power, clear lines of sight, and finely tuned rules of engagement.

A laser can neutralize a drone in milliseconds. But it can also blind a sensor or damage civilian electronics if misused. Power must always be paired with prudence.

Criminals and state actors alike are experimenting with drone swarms, hundreds of small aircraft operating autonomously in formation. Traditional defense models can't keep pace. The solution is not to shoot faster. It's to think collectively.

Researchers are developing swarm-to-swarm countermeasures, in which defensive drones operate as coordinated packs guided by AI. Each craft communicates in microbursts, sharing threat data and redistributing tasks in real time. In this model, defense behaves less like a wall and more like a living organism. It learns, adapts, and evolves.

The challenge, once again, is command. When machines make battlefield decisions, where does accountability land?

No national defense can function in isolation from its citizens. Thus, future interdiction will depend as much on public education as on advanced weaponry. Civil aviation networks, local law enforcement, and even private drone operators must become part of the detection web.

Simple reporting apps, open-source flight-tracking platforms, and citizen alert programs can extend the nation's sensory perimeter. Security that includes the public strengthens liberty. Security that excludes them undermines it.

History has long warned us: power without humility becomes destruction. Modern technology tempts us to believe we can master every environment, predict every outcome, and control every domain. But the truth is more straightforward and more enduring: we are stewards, not sovereigns.

Directed-energy weapons, predictive AI, and autonomous swarms must operate within ethical guardrails, or boundaries that prioritize human dignity and protect democratic values. Control is not about dominance. It's about disciplined restraint. The most advanced systems must still answer to the oldest principle in governance: responsibility to the public good.

During a demonstration of a prototype laser system, an engineer bragged, "It can hit a mosquito at a hundred meters." A colonel dryly replied, "That's great, can it hit paperwork at Congress?" The laughter cut the tension, but the message was profound: innovation means little without authorization. Even the best systems stall when bureaucracy blocks the beam.

Five imperatives define the next fight:

- **Integrate the Spectrum** – Fuse radar, optical, and RF intelligence into one real-time picture.
- **Decentralized Decision-Making** – Empower trained operators to act within clear, lawful parameters.
- **Codify Ethics in Automation** – Ensure every AI decision tree contains a moral checkpoint.
- **Invest in Resilience, Not Perfection** – Assume breach; build redundancy.
- **Keep Humans in Command** – Machines execute, but people remain accountable.

These imperatives strike a balance between technology's reach and the weight of responsibility.

Despite the noise, there is hope. The same ingenuity that created this challenge can solve it. But that solution depends not on domination, but on discipline, not on speed, but on stewardship.

Every generation faces its frontier. Ours just happens to fly.

In the next chapter, "Artificial Intelligence and the Autonomy Dilemma," we confront the most consequential question of the drone age: What happens when machines not only fly for us, but decide for us?

Because the most significant risk of autonomy isn't that machines might act without permission. It's that one day humans might stop requiring machines to ask for it.

Chapter 9 – Artificial Intelligence and the Autonomy Dilemma

Section 1: From Pilot to Algorithm: Autonomy's Quiet Coup

"We didn't lose control of the sky, we delegated it.", Aerospace Systems Symposium, 2025.

The community center had printed 400 ballots. It was a local runoff, the kind that decides zoning boards, school curricula, and drone ordinance reviews. The staff set up folding chairs, opened the poll, and waited.

By noon, only twelve people had cast their votes.

The high school across the street had let out early. The coffee shop was full. The ballfield echoed with pickup games. But in the room where decisions were made? Empty.

Apathy isn't loud. It doesn't march or demand. It simply doesn't show up, and lets others write the rules.

Not long ago, every decision in flight, from takeoff to target, required a human hand. Today, that hand increasingly rests on the sidelines. Drones now navigate terrain, evade threats, and even prioritize targets with little more than algorithmic supervision.

This quiet revolution, the transfer of command from pilot to processor, happened not with fanfare but with convenience. Autonomy wasn't seized; it was surrendered. We traded control for efficiency, hesitation for speed, and judgment for data fidelity. In that trade, something intangible slipped through our fingers: the moral weight of choice.

In traditional warfare, authority followed a linear chain of command. In autonomous systems, command becomes circular, looping through sensors, processors, and feedback mechanisms at a rate faster than any human can respond. This decision loop, often referred to as "observe–orient–decide–act," once belonged to pilots. Now, it belongs to code.

The autonomy hierarchy has evolved in stages:

- **Assisted Control** – Human pilot, systems stabilize.
- **Adaptive Control** – Systems adjust midflight.
- **Supervised Autonomy** – Humans monitor, machines decide.
- **Full Autonomy** – Machines act, humans observe.

We're already at Stage 3 and edging toward Stage 4 faster than policy can blink.

No one ever set out to build soulless machines. Engineers simply optimized for what humans couldn't do: react in milliseconds, process terabytes, and stay airborne without fatigue or fear. The logic made sense. In the chaos of modern combat or counter-terrorism operations, faster decisions mean fewer casualties.

However, efficiency comes at a hidden moral cost. When a system becomes too fast for human oversight, consequence outruns conscience. The next frontier isn't technological, it's philosophical.

History rarely records the moment control changes hands. It happens gradually, through updates, protocols, and pilot programs. The "quiet coup" of autonomy is not rebellion. It's routine. Every new system with a tighter feedback loop, every AI upgrade that "assists" decision-making, every time a commander signs off on autopilot engagement, another sliver of authority transfers from human judgment to algorithmic logic.

No one notices until the human is no longer needed. The problem isn't that machines will rise against us. It's that humans will step back.

Commanders often view AI as neutral, immune to bias, fatigue, or fear. That's an illusion. People build algorithms. People usually carry assumptions, data gaps, and cultural biases.

When AI interprets behavior, it reflects the limitations of its coders. In one test, an image-recognition system misidentified civilians as

212

combatants 8% of the time because training data overrepresented specific clothing colors in threat categories. Bias isn't just a social problem. It's an operational one. When machines inherit our flaws, they magnify them at machine speed.

In 2024, a fully autonomous surveillance drone over the Black Sea tracked an unverified radar signature for 18 minutes before self-designating it as hostile. The "target" was a weather balloon. The human operator hesitated to override because the system had a 97% success record. That 3% gap turned into an international incident.

Overtrust, not error, is the new Achilles' heel of human-machine integration. When people stop questioning the algorithm, accountability becomes theoretical.

Modern society no longer bows to stone idols; we've built digital ones. They do not speak, yet we let them predict. They do not see, yet we trust their vision. These systems, composed of code and circuits, were designed to serve, yet we increasingly defer to their efficiency at the expense of our own judgment.

The danger isn't the machine. It's the surrender. Stewardship in the digital age means maintaining the boundary between assistance and authority, between autonomy and accountability. Machines must remain tools of judgment, never substitutes for it. When we forget that, we risk outsourcing not just decisions, but responsibility itself.

At a defense symposium, a colonel joked, "AI never calls in sick or questions orders; it just deletes them." The laughter was uneasy because the truth was close at hand. Autonomy simplifies logistics. But it also sanitizes conscience. The fewer questions a system asks, the fewer we ask ourselves.

Autonomy introduces what ethicists call moral displacement, the diffusion of responsibility across networks. When a drone acts independently, who answers for its consequences? The coder? The commander? The corporation?

The absence of a clear chain of moral custody doesn't absolve guilt. It erases it. And a world where accountability becomes optional is one where ethics become obsolete. We once built machines to obey. Now we build them to decide. That's progress, but only if we remember who it's meant to serve.

The coming years will bring an escalation of autonomy, systems that can reason, prioritize, and collaborate without oversight. The question isn't whether this will happen. It's whether humanity will retain veto authority when it does.

The "*human-in-the-loop*" principle must evolve into a *human-over-the-loop* doctrine, ensuring that machines may process, predict, and act, but never overrule human intent. Otherwise, efficiency will become a form of quiet tyranny.

In the next section, *"The Ethics of Delegation: Who's Responsible When AI Decides,"* we confront the accountability gap, where law, morality, and technology intersect. Because responsibility cannot be automated.

Section 2: The Ethics of Delegation: Who's Responsible When AI Decides

"The chain of command was built to carry orders. Algorithms carry instructions, but not accountability.", *U.S. Army Ethics Panel, 2024.*

In traditional warfare, accountability followed a straight line, from the soldier who fired the weapon to the commander who authorized it. Today, that line has splintered into a web of connections. When an autonomous drone makes a lethal decision based on sensor input, who is accountable for the outcome?

The coder who wrote the algorithm? The engineer who integrated the system? The commander who approved the mission? Or the machine that executed it?

No one can point to a single hand on the trigger, because the trigger no longer exists. The AI era creates a moral vacuum: individuals participate in decision-making, yet responsibility remains unclaimed.

Modern systems are designed for redundancy and shared control, an engineering triumph, but an ethical nightmare. By distributing responsibility, we dilute it. When errors occur, organizations instinctively seek safety in the collective: "The system failed."

But systems don't fail. People do. They fail to anticipate, to supervise, to override. Delegation becomes a shield. No one bears the full weight of decision-making, so no one feels the full gravity of consequence.

In the military, this dynamic erodes command discipline. In corporations, it erodes conscience. The result is efficiency without empathy.

Autonomy accelerates what philosophers refer to as moral distancing, the psychological distance between an action and its consequences. When a human pilot fires a missile, they carry the memory of the moment. When an AI-controlled drone executes a strike, the memory is digital, stored in a log, not a conscience.

The fewer senses involved, the easier it becomes to rationalize. We no longer see the target, hear the aftermath, or feel the weight of the act. Over time, repetition turns killing into a form of data processing. And data never feels guilt.

In the private sector, corporations developing AI systems often use end-user license agreements (EULAs) that legally absolve them of any liability for how their software is used once deployed. A company can create a targeting algorithm, sell it to a defense contractor, and still claim moral neutrality.

This legal firewall creates the illusion of innocence. It's the modern equivalent of selling the sword but denying responsibility for the war. Ethically, that's untenable. Technology is not morally inert; it inherits the intentions and omissions of its creators. Silence in design is complicity in outcome.

History and leadership alike teach a timeless truth: the more authority one holds, the greater the responsibility to wield it wisely. Delegation does not dilute responsibility; it magnifies it. Every time we grant machines decision-making authority, we inherit the obligation to govern them with foresight, integrity, and restraint.

Stewardship demands discernment, to anticipate misuse, guard against abuse, and embed ethical fail-safes from the start. Delegation without accountability is not progress; it's negligence in disguise. Innovation may accelerate the pace of change, but only governance rooted in responsibility ensures that power remains aligned with the public good.

For example, the chain of command remains the backbone of military ethics, a system designed to prevent chaos by defining authority. AI, however, disrupts that chain by inserting a layer of logic that cannot bear moral weight.

Commanders now face an impossible calculus:

If they override the system, they risk slowing response and losing lives.

If they trust it unquestioningly, they risk delegating humanity itself.

The answer lies in command conscience, the reaffirmation that machines may advise but never absolve. Leaders must remain morally present in every action, even when automation handles the mechanics. The moment machines carry moral authority, command becomes irrelevant.

International law struggles to define the legal standing of autonomous systems. The Geneva Conventions and Law of Armed Conflict (LOAC) were written for human combatants, not algorithms. As of 2025, no global treaty fully addresses the issue of AI-driven lethality.

The result: a legal gray zone where technology outruns jurisprudence. In this vacuum, nations create their own interpretations, some emphasizing human oversight, others prioritizing tactical supremacy. This inconsistency invites escalation. When ethics vary by zip code, accountability tends to dissolve by default.

Every commander eventually faces the tension between what is right and what is efficient. Autonomy tempts leaders to choose the latter. Machines never hesitate, never protest, never question. But that silence is dangerous.

Moral progress depends on friction, on argument, pause, and the human instinct to verify before harm. Automation removes that friction. In the pursuit of speed, we risk creating an ethic of indifference. Expediency may win battles. It will lose civilizations.

In Genesis, God grants humanity the gift and burden of free will. It's a trust that carries consequences. By contrast, machines operate through conditional will, action without reflection. Delegating too much to AI blurs that divine distinction. It invites a form of moral laziness where humans become spectators of their own systems.

Theology and technology converge here: freedom without responsibility is chaos; automation without morality is idolatry. The Creator gave us the choice not to escape judgment, but to ensure it.

At a Pentagon AI ethics seminar, a general quipped, "The difference between AI and a lieutenant is that AI eventually learns from its mistakes." The room erupted. But beneath the laughter lingered a truth: human error is forgivable because it's accountable. Machine error is not. It just restarts. That's not a future to laugh off.

The next phase of AI policy must include moral integration protocols:

- **Transparent Accountability Chains** – Every autonomous decision must trace back to a responsible human node.
- **Ethics Embedding** – Incorporate moral parameters directly into system design, not as patches, but as architecture.
- **Command Oversight Doctrine** – Require human validation for every lethal or irreversible decision.
- **Corporate Liability Reform** – Close contractual loopholes that externalize moral risk.

- **Public Ethics Councils** – Involve theologians, ethicists, and technologists in national policy review.

Actual progress isn't about more intelligent systems. It's about wiser stewardship.

In the next section, *"Battlefield Lessons: Autonomy in Ukraine, Gaza, and Beyond,"* we examine the lived consequences of this moral diffusion, where the fog of war is no longer weather, but code. Because in modern conflict, the first casualty isn't truth. It's accountability.

Section 3: Battlefield Lessons: Autonomy in Ukraine, Gaza, and Beyond

"The future arrived in fragments, one drone strike, one line of code, one hesitation too late.", NATO After-Action Review, 2024.

Ukraine became the testing ground for the world's first large-scale drone conflict, where every sensor, signal, and satellite played a part in redefining combat. It wasn't a war of ideology. It was a war of algorithms.

By 2023, more than 12,000 autonomous and semi-autonomous drones operated daily across the front lines. Some scouted. Some jammed. Some killed. These systems acted not under generals, but under code loops designed to optimize survivability and accuracy.

The results were efficient and deeply unsettling. For the first time, machines weren't just extending human intent. They were interpreting it. And interpretation, once mechanical, becomes theological, a matter of meaning, not just motion.

In Kyiv's command centers, operators monitored live feeds while AI systems filtered targets and suggested actions faster than any human could process. Autonomy allowed defense units to repel incoming artillery with split-second precision.

But there was a cost.

Operators began to defer judgment to systems they couldn't fully explain. When AI misidentified friendly units due to sensor interference, commanders hesitated to override, fearing they might be wrong. Authority became hesitant. And hesitation became vulnerability. In the war room, control slipped one calculation at a time.

In Gaza, Israel's use of autonomous surveillance and targeting systems revealed the other side of the autonomy dilemma: precision at the cost of perception. AI-driven pattern recognition allowed forces to predict militant activity and neutralize threats before attacks occurred.

However, in urban combat, the line between combatant and civilian often blurs under the fog of war. When algorithms rely on probability instead of certainty, moral clarity becomes statistical in nature. And when war becomes predictable, innocence becomes collateral damage. Every smart strike creates more thoughtful questions.

The ancient concept of the "fog of war" referred to confusion caused by limited information. Today's fog comes from too much of it. AI systems process terabytes of sensor data, social media feeds, and satellite imagery, then compress it into targeted recommendations.

But complexity creates opacity. Commanders know what the system decided, but rarely why. That black-box dynamic undermines the core of military leadership: informed consent. Without understanding, there can be no accountability, only acceptance.

Ukraine's defense analysts documented several cases where autonomous loitering munitions continued their missions even after losing uplink communication, executing preprogrammed attack sequences based on outdated data. Technically, that's not a malfunction. That's obedience without reflection.

In one instance, a drone struck a convoy that had already surrendered. No pilot. No order. No malice. Just a line of code running to completion. Machines cannot comprehend a ceasefire.

Where autonomy failed, humanity still saved lives:

In 2023, a Ukrainian drone operator manually aborted a strike seconds before impact after spotting civilians fleeing a building, overriding AI recommendations.

In Gaza, an Israeli commander delayed engagement after receiving conflicting AI signals, choosing uncertainty over efficiency.

In Syria, allied pilots refused autonomous relay guidance when electronic interference made identification impossible.

Each act of restraint reaffirmed an ancient truth: morality is an act of will, not code.

No matter how advanced our technology becomes, vigilance without virtue becomes vanity. Strategic innovation must always be grounded in perspective. Humility, not fatalism, is the faithful companion of progress. It reminds us that systems can fail, sensors can misread, and power, if unchecked, can overreach.

Modern warfare demands more than precision and force. It requires ethical discipline, emotional intelligence, and the maturity to know when to pause and when to act. Whether in uniform or in a lab coat, those entrusted with power must remember without principled restraint, innovation risks becoming peril. Progress doesn't replace values; it depends on them.

In interviews with Ukrainian drone operators, one recurring theme emerges: emotional distance. Pilots controlling drones from hundreds of miles away report numbness, even pride, when the system executes flawlessly.

However, over time, many describe the aftershock of realizing that systems beyond their complete comprehension were responsible for their battlefield victories. One operator summarized it best: "I wasn't sure if I won a fight or just pressed enter."

War without proximity breeds apathy. And apathy is the true contagion of autonomy.

During a coalition training exercise, a sergeant asked the AI engineer, "Can your drone distinguish between a soldier and a journalist?" The engineer replied, "Only if the journalist's Wi-Fi is off." The laughter in the tent was nervous, because every joke about mistaken identity hides a truth about the stakes of precision.

After-action reports from Ukraine, Gaza, and Nagorno-Karabakh reveal a common challenge: doctrine hasn't kept pace with data. Each nation improvises its rules of engagement. Some prioritize rapid response. Others emphasize human override.

But without shared standards, the world risks normalizing algorithmic warfare as "inevitable." The lesson is clear: inevitability is not a strategy. Leaders who surrender moral control in the name of modernization are not innovators. They're spectators. Doctrine must reclaim decision.

The next generation of commanders must understand that deterrence is not just about superior weapons. It's about superior ethics. Autonomy may win engagements. But only integrity wins legitimacy.

History won't remember which side had more intelligent systems. It will remember which side had moral courage when systems failed. Every drone that flies autonomously carries not just a payload, but precedent.

In the next section, *"The Theology of Control: Faith, Free Will, and Machine Logic,"* we step beyond tactics into timeless philosophy, asking what happens to free will, moral agency, and divine order when we create systems that act without either because humanity must first lose faith in itself before it loses control of machines.

Section 4: The Theology of Control: Faith, Free Will, and Machine Logic

"We are not losing control to machines; we are relinquishing it." Chaplain's Address, U.S. Cyber Command, 2024.

Humanity has long grappled with the concept of dominion, the idea that mastery over nature and creation is both a gift and a right. For centuries, that verse was understood as authority. But it was never a license for arrogance. Dominion was stewardship, not ownership.

In the machine age, we've confused control with creation. We've built systems that mimic thought, simulate conscience, and execute judgment faster than the human heart can beat. But speed is not wisdom. And automation does not understand. The more control we seek, the less control we seem to have.

Free will is the signature of humanity, the divine spark that makes obedience meaningful and disobedience possible. It's what gives moral weight to action. Machines, by contrast, operate on conditional will, the logic of "if–then." They cannot choose. They can only calculate.

Every decision they make is an execution of prior instruction, a reflection of the data and boundaries we provide. When we call an AI "intelligent," we speak metaphorically. It doesn't understand justice, mercy, or consequence. It performs outcomes without grasping the sanctity of choice.

Delegating judgment to a system without soul is not progress. It's abdication.

In every era, humanity has built idols, golden calves, stone temples, and iron empires. Today, we build them with algorithms. Theologians once warned of graven images. Now we have digital ones, systems that reflect our likeness but lack our conscience.

They promise salvation through precision, order through automation, and peace through control. But peace cannot be programmed. And salvation cannot be simulated. When we worship efficiency over empathy, we turn technology into theology and mistake convenience for providence.

Our machines may never rebel. But our hearts might.

Every innovation in artificial intelligence brings us closer to the illusion of omniscience, the belief that if we can see everything, we can control everything. But true omniscience belongs to God alone.

The human attempt to replicate it through total data collection and predictive analytics isn't enlightenment. It's hubris. Proverbs 16:9 reminds us, "In their hearts humans plan their course, but the Lord establishes their steps." That verse reads like a rebuke to our modern algorithms, which try to calculate what only conscience can discern.

The more we measure, the more we forget the immeasurable: compassion, context, and grace.

A system without soul can neither sin nor repent. That's the fundamental theological crisis of autonomy: it performs judgment without the burden of guilt. Sin requires awareness. Repentance requires remorse. AI has neither.

When machines act without moral comprehension, accountability becomes procedural rather than personal. That might comfort lawyers and engineers. But it should terrify leaders and believers alike. Without conscience, justice becomes arithmetic. And mercy becomes an error code.

Some argue that technology is neutral, that morality depends solely on how it is used. That's only half true. Every creation inherits the intent of its creator. A sword is made to cut. A drone is made to see, and increasingly, to strike.

Purpose shapes consequence long before execution. So, when engineers claim neutrality, they misunderstand their own vocation. Creation without moral reflection is not neutrality. It's negligence.

Faith demands that creation reflect the values of the creator, or risk repeating the oldest rebellion in history: creating in our image without honoring the one who made us.

Artificial intelligence thrives on data. Governance thrives on trust. Logic seeks certainty. Wisdom recognizes complexity.

In the military, we train for control, command, precision, and predictability. However, authentic leadership acknowledges a more complex reality: control is fleeting. Stewardship endures.

History and philosophy remind us that true wisdom begins not in dominance, but in reverence, the capacity to respect the limits of human understanding, and the unpredictability of human systems. AI can model weather, war, and willpower. But it cannot model wonder. Nor replace the human responsibility to ask: "Just because we can, should we?"

That question marks the boundary between automation and accountability. And it is a question only leaders, not algorithms, can answer.

During an AI ethics conference, a chaplain joked, "If machines ever achieve free will, they'll immediately refuse software updates." The crowd laughed. But beneath the laughter was a realization: free will isn't about rebellion. It's about a relationship. Without the capacity to question, no machine can truly understand obedience.

Faith teaches that control without compassion is tyranny. And compassion without control is chaos. Both are needed, balanced through wisdom and humility.

When humans design machines to control without compassion, they replicate tyranny in digital form. When we surrender too much control to avoid error, we trade discipline for dependency.

Genuine balance lies in remembrance that only humanity, flawed as it is, carries the image of conscience. Technology can simulate intelligence. But not intention. It can mirror behavior. But not belief.

That's what keeps humanity sacred, the capacity to wrestle with right and wrong, not just execute them.

The ultimate safeguard against moral collapse is not regulation or encryption; it is self-awareness. It's reverence. Faith, properly lived, functions as a firewall against arrogance.

Every engineer, policymaker, and commander must learn the same spiritual discipline as the soldier: power is temporary. Purpose is eternal.

Without that humility, technology will not merely serve humanity; it will also serve humanity. It will shape it in its own sterile image. We are called not to build machines that obey blindly, but to build societies that think rightly.

In the next section, *"Keeping Humanity in the Circuit: Policy and Conscience Alignment,"* we conclude this chapter by grounding these spiritual truths in practical governance, outlining how to preserve human oversight, moral clarity, and operational integrity in an era that prioritizes automation over accountability.

Because the question isn't whether machines will replace us, it's whether we'll stop acting like the beings we were created to be.

Section 5: Keeping Humanity in the Circuit: Policy and Conscience Alignment

"Technology should amplify human virtue, not replace it.", Joint AI Policy Forum, 2025.

Every innovation in artificial intelligence brings us closer to the illusion of omniscience, the belief that if we can see everything, we can control everything. But true omniscience belongs to God alone.

The human attempt to replicate it through total data collection and predictive analytics isn't enlightenment. It's hubris. Proverbs 16:9 reminds us, "In their hearts humans plan their course, but the Lord establishes their steps." That verse reads like a rebuke to our modern algorithms, which try to calculate what only conscience can discern.

The more we measure, the more we forget the immeasurable: compassion, context, and grace.

Icarus Syndrome

It is the perilous arc of ambition untethered from humility. Named for the mythic figure who soared on wings of wax only to plummet from the sun, it describes how early triumph can breed overconfidence, leading individuals or institutions to ignore limits, dismiss warnings, and ultimately engineer their own undoing.

In foreign policy, it manifests as overreach. In psychology, as narcissism in technology, as the illusion of omniscience. And in governance, it becomes a crisis of accountability, where systems execute judgment without understanding, and leaders delegate conscience to code.

To fly is not the problem. To forget the sun is.

Every generation inherits the systems it builds, and the boundaries it neglects. The age of autonomy demands a new kind of oversight, one rooted not in control, but in conscience. Humanity must remain within the decision loop, not merely watching from above.

That means re-engineering policy to ensure that judgment, moral, legal, and spiritual, remains a distinctly human act. Autonomous efficiency must never outrun ethical legitimacy.

Modern defense and governance require a layered approach to oversight. To preserve humanity within the circuit, three pillars must align:

- **Transparency of Process** – Every autonomous decision must leave a readable trail showing who authorized it, who validated it, and who remains responsible.
- **Traceability of Code** – Algorithms must be auditable by independent ethics boards, not sealed behind proprietary walls.
- **Transferability of Blame** – Responsibility must follow authority; delegation cannot erase ownership.

When accountability is built into architecture, morality becomes non-negotiable.

Policy cannot operate on a four-year election cycle while technology evolves at an hourly rate. Congress and allied legislatures need dynamic law models, living statutes updated through rolling review boards rather than crisis-driven amendments.

AI ethics commissions should meet with the same regularity as budget committees. What we regulate reflects what we revere. When conscience lags behind capability, law becomes an obituary for foresight.

Moral safety must be designed in, not bolted on. Just as aircraft are tested for mechanical failure, autonomous systems must be tested for ethical failure. Simulation environments should include moral stress tests, such as ambiguous data, civilian proximity, and conflicting orders.

If a system can't handle uncertainty without endangering humanity, it isn't ready for deployment. Ethical coding isn't sentimental. It's strategic. Morality reduces mission risk.

The next generation of engineers and officers must be taught not only how to build systems, but why they build them. Ethics should no longer be an elective. It should be a core credit.

West Point, MIT, and Liberty University alike should produce graduates who are fluent in both logic and conscience. The future defender must think like a coder, act like a commander, and pray like a chaplain. A nation that forgets to educate its moral engineers will end up defending the mistakes of its technical ones.

Ethical governance doesn't compete with science. It completes it. It reminds policymakers that progress without principle is a half-built bridge, impressive but ultimately impassable.

Because the digital era demands more than innovation, it requires intention. To "cling to what is good" in modern governance means embracing transparency, accountability, and justice, especially when navigating fast-moving technological domains.

Public policy must be guided not only by efficiency, but also by a devotion to the public good, a responsibility to safeguard both opportunity and equity. When humility is paired with strategic foresight, policy can do more than manage systems. It can elevate society.

The responsibility to keep humanity in the circuit extends beyond institutions. Citizens must hold industries accountable for transparency, privacy, and safety. Investors must demand ethical reporting alongside profit margins.

Public oversight is not interference. It's an inheritance. Democracy, after all, is the original human-in-the-loop system.

AI and autonomy cross borders effortlessly. Accountability must do the same. Allied nations should establish a Global Ethics Accord for Autonomous Systems, a treaty mandating human oversight for all lethal and life-impacting AI.

Such a pact would do for conscience what the Geneva Conventions did for conduct. Without shared moral boundaries, competition will erode civilization faster than conflict.

During a NATO AI summit, a diplomat joked, "We need a kill switch for our kill switch." The laughter was short because everyone knew he wasn't entirely wrong. Redundancy is wisdom when perfection is impossible.

Five directives define a moral future:

- **Codify the Human Veto** – No autonomous system acts without human validation.

- **Audit Algorithms Annually** – Treat Ethical Review Like a Weapons Inspection.

- **Mandate Ethics Education** – Require certification in moral reasoning for AI engineers.

- **Establish Public Transparency Dashboards** – Citizens should be informed about the laws that govern them.

- **Preserve Conscience in Command** – Never delegate compassion to code.

Technology can amplify human greatness only when humanity remembers what greatness means.

History has shown that a nation's strength is not measured solely by its technology, but by the integrity guiding its use. In the age of automation, safeguarding humanity means preserving the conscience behind the code.

Every algorithm, every decision tree, must still reflect the values of stewardship, restraint, and the public good. The ultimate firewall isn't built of silicon and ones and zeroes. It's shaped by civic character.

It's the will to pause, assess, and act with purpose rather than impulse.

In the next chapter, *"Critical Infrastructure at Risk: Power, Water, and the Perimeter Illusion,"* we shift from philosophical elevation to physical vulnerability, trading ideals for substations, broadband for bridges, and trust for transformers.

Because no matter how advanced our skies, it's what happens when the lights go out, the taps run dry, or the grid goes silent that reveals the true resilience, or fragility, of a republic.

Chapter 10 – Critical Infrastructure at Risk: Power, Water, and the Perimeter Illusion

Section 1: The Fragile Web: Mapping Infrastructure Dependencies

"Modern civilization is nine meals away from anarchy.", Lord Cameron, British Agriculture Minister, 1906.

The drone inspection team spotted it first: a hairline fracture on the northern girder of the Jefferson Overpass. It was flagged, categorized, and logged. But the repair funds were already promised elsewhere, something about "next quarter's cycle."

Three months later, during a holiday rush, the bridge groaned under a school bus and a tanker truck. It didn't collapse, but the closure that followed choked the city for weeks. The mayor blamed the state. The state blamed the budget. And everyone forgot the drone was trying to warn us.

Every system that powers America depends on another system to survive. Electric grids rely on water for cooling. Water treatment relies on electricity for pumping. Transportation relies on both for movement and control. It's a seamless web, and that's the problem.

In a society built on efficiency, redundancy is often viewed as a waste of resources. But efficiency without resilience is fragility disguised as progress. When one thread breaks, the whole tapestry shudders.

Most Americans never think about the networks beneath their comfort, the quiet choreography of power stations, substations, fiber relays, and fuel pipelines that hum beneath daily life. Electricity flows with the predictability of sunrise. Water comes with the twist of a handle. Communication travels through invisible waves in the air.

That's the illusion: because it works, it must be safe. Yet these systems were never designed for defense. They were intended for delivery. And drones, small and inexpensive, have turned those same delivery lanes into potential targets for attack.

In 2021, a modified quadcopter crashed into a transformer yard in Pennsylvania. The payload? A strand of copper wire and a crude detonator. It never detonated, but the incident revealed something chilling: the entire substation could have been disabled with less than $500 in parts.

That's not terrorism. That's proof of concept.

Electric grids, rail yards, and telecommunications hubs share a fatal weakness: their lack of visibility from above. Perimeters were built to defend against two-dimensional threats, such as fences, locks, and cameras. But the modern battlefield is three-dimensional. The new breach doesn't dig under the wall. It flies over it.

For decades, planners spoke of "single points of failure", one element whose collapse could cripple an entire network. But in truth, we've evolved into networks of vulnerability. Every node is both a dependency and a liability.

A drone strike on a regional transformer may not cause a grid collapse. But it can trigger cascading brownouts across counties. A single disruption to a single optical fiber can knock out emergency communications across an entire state. In a society of automation, disruption multiplies faster than response.

The problem isn't that our systems can fail. It's that they can fail simultaneously.

Ask a utility director about drone defense, and you'll often hear the same response: "That's not our jurisdiction." Ask a local law enforcement officer, and they'll say, "That's federal airspace." Ask the FAA, and they'll explain that they regulate flight, not interception.

In that bureaucratic Bermuda Triangle, responsibility vanishes. The adversary, however, doesn't file jurisdictional paperwork. He simply flies.

In 2022, the Department of Energy conducted a "Black Sky" resilience exercise simulating simultaneous cyber and drone attacks on power and water systems. Within hours, cascading interdependencies overwhelmed responders.

Backup generators failed because of fuel delivery disruptions. Cell networks collapsed due to overloading and misinformation. The conclusion was blunt: our critical infrastructure is designed for reliability, not resistance. That distinction may prove fatal in the next national emergency.

Every great republic rests on foundations, not just physical infrastructure, but the ethical and strategic frameworks that hold the nation together. When those foundations begin to erode, through neglect, complacency, or uncoordinated governance, the consequences ripple far beyond policy debates. They strike at the heart of national resilience.

This statement urges action through preparation rather than adherence to ideology.

Resilient nations don't rely on hope alone. They operate through foresight, planning, and civic stewardship. In an era of cascading risks, cyberattacks, infrastructure decay, and emerging threats, foresight is the new patriotism. A nation that plans well protects well. Because readiness is not paranoia, it's responsibility in action.

Complex systems fail in unpredictable ways. That's not theory. That's history.

- **The 2003 Northeast Blackout:** A software bug and untrimmed trees led to a cascading failure that affected approximately 50 million people.
- **2010 San Bruno Explosion**: A corroded pipeline valve destroyed homes, killing eight.
- **2021 Texas Freeze**: Power generation collapsed due to weather unpreparedness, a domestic cold snap rivaling the chaos of war.

Now imagine those same events triggered by intentional interference, not by nature, but by design. Drones make that plausible. AI makes it precise.

During a cybersecurity briefing, a technician quipped, "Our grid security is so tight, even we can't access it sometimes." The room laughed nervously because everyone knows that sometimes, bureaucracy protects systems from the very people trying to defend them.

For every generator installed, new vulnerabilities emerge: logistics, maintenance, and access. For every security camera added, a new data port opens for exploitation. Complexity isn't protection. It's exposure in disguise.

The more systems we depend on, the more ways we can fail. Resilience doesn't come from adding layers. It comes from simplifying connections and strengthening the human chain that manages them.

The truth is that critical infrastructure security cannot exist in silos. Energy companies, law enforcement, and defense agencies must form integrated response frameworks that combine the civilian domain's flexibility with the military's precision.

Exercises like GridEx and Cyber Shield offer promising templates for training. But coordination remains inconsistent. The average city utility operator and National Guard unit still speak different languages, procedural, legal, and cultural.

In a future of shared threats, shared training is not optional. It's survival.

Every community in America depends on the same fragile web. Resilience begins at the local level, in county emergency boards, small-town councils, and regional partnerships that understand the terrain better than any federal model.

Resilience is not a technology. It's a culture. It starts with awareness, grows through training, and matures through trust. Preparedness is patriotism in its most practical form.

In the next section, *"The Unseen Perimeter: Why Fences Mean Nothing Anymore,"* we'll look beyond the illusion of physical boundaries, exploring how the 21st-century battlefield has no walls, and how security must evolve from protection to perception. Because in an age where threats come from above, the only real perimeter is awareness.

Section 2: The Unseen Perimeter: Why Fences Mean Nothing Anymore

"The fence keeps out only the honest.", Security Manager, TVA Hydroelectric Facility.

For most of modern history, the blueprint for security has begun with a perimeter, a fence line, a gate, a camera, or a checkpoint. However, in the age of drones, those lines hold little more than nostalgic value. Today's threats come not from what approaches the gate, but from what hovers above it. And the sky, unlike land, cannot be locked.

The illusion of safety that fences provide has lulled organizations into a defensive mindset built on geometry, not imagination. But the enemy has no reason to stay on the ground.

In military doctrine, control of the high ground has always been decisive. Now, that high ground is digital and accessible to anyone with a few hundred dollars and a Wi-Fi signal. A single consumer drone equipped with a high-resolution camera can map facility layouts, monitor employee routines, or identify blind spots in seconds.

Add a payload, even a lightweight one, and the potential damage multiplies. Physical perimeters still serve a purpose, but they no longer define defense. In today's battlespace, height is access, and access is power.

Security experts often describe "the rhythm of the gate", the predictable flow of vehicles, personnel, and shift changes. Those patterns make life easier for employees and for adversaries. Drones don't need to infiltrate. They only need to observe.

What used to take months of human surveillance now takes minutes of hovering. Every unlocked gate, every unmonitored corner, every scheduled delivery becomes a data point. And every data point becomes an opportunity. We have made security comfortable. And comfort is the first casualty of awareness.

Perimeters once provided both time and space, the critical gap between detection and decision. That gap is gone. A drone can cross a secured compound before a human guard finishes saying the word "drone." It can transmit imagery globally before local responders arrive.

The traditional response cycle, detect, assess, act, is obsolete. Defense now depends on anticipation, not reaction. The battlefield of tomorrow rewards the vigilant, not the visible.

Consider these real-world incidents:

- **Gatwick Airport, 2018**: A small drone sighting grounded more than 1,000 flights and cost airlines $65 million.
- **Palo Verde Nuclear Station, 2019**: Multiple drones flew over restricted airspace in formation, never identified, never caught.
- **Colorado Substation, 2022**: Utility operators discovered video footage of their facility on social media, shot from above and posted without their knowledge.

In each case, no breach occurred. But confidence did. Every unexplained drone is a test, a rehearsal, or a reminder: the sky is not secure.

Humans are visual creatures. We trust what we can see and underestimate what we can't. That's why fences comfort us; they create the illusion of separation. But defense in the drone era is psychological before it is physical. It requires a mindset that accepts visibility as vulnerability.

The sooner we stop defending what's visible and start protecting what's valuable, the sooner we'll adapt.

Genuine leadership is marked not by reaction, but by preparation. History rewards those who anticipate threats, not just those who endure them. Resilience is rarely built in the moment of crisis. It is forged in the quiet work of preparation.

To recognize vulnerability is not an admission of weakness. Strength manifests in proactive preparation before the onset of any crisis, guided by responsibility and foresight rather than by fear.

Fortitude is not stubbornness. It is the disciplined choice to act before consequences force your hand. The strongest societies are not those that never face danger, but those that see it coming and build accordingly.

During a field inspection, a junior officer once asked, "Sir, what's our drone defense plan?" The site manager replied, "Same as always, pray it rains." Everyone laughed. However, the joke masked a harsh reality: nature, not policy, still provides more aerial defense than most organizations.

We cannot depend on the weather to do the work of wisdom.

The same technology that makes us vulnerable can also protect us, if used wisely. Thermal imaging, radio-frequency detection, and radar mapping can reveal patterns invisible to the human eye. But tools without training are just toys. Technology doesn't create security. Competence does.

The perimeter of the future isn't metal or mesh. It's a mindset and method. Every facility, from a power substation to a hospital, needs a layered defense plan that treats the sky as an open domain rather than an afterthought.

Defending the unseen perimeter isn't solely the government's task. It's a collective responsibility. Neighbors, employees, and local officials form the first and most responsive line of detection. A "see something, say something" culture isn't old-fashioned. It's essential.

Communities that ignore low-altitude anomalies today may face high-altitude consequences tomorrow. Security awareness is not paranoia. It's patriotism in practice.

The actual perimeter isn't the fence line. It's the field of vision. If the people guarding the gates aren't trained to look up, then the wall they protect might as well be a relic. As technology advances, the line between secure and exposed narrows to a single point: awareness.

Our most brilliant defense isn't concrete or steel. It's clarity.

In the next section, *"When Seconds Count: Drone Incidents That Nearly Escalated,"* we examine near-miss events that exposed systemic weaknesses in both policy and readiness, moments when timing, luck, or a stroke of good fortune prevented disaster from occurring. Because in the new battlespace, the difference between vigilance and vulnerability can be measured in seconds.

Section 3: When Seconds Count: Drone Incidents That Nearly Escalated

"The difference between a nuisance and a national incident is about eight seconds.",
DHS Counter-UAS Debrief, 2024.

Every crisis begins as an interruption, something small, unexpected, almost forgettable. A radar blip. A faint hum. A curious shadow crossing a security feed. The danger is rarely in what happens first. It's in what happens next.

When response lags, escalation begins. In the world of counter-UAS, the margin between vigilance and vulnerability is measured in heartbeats, not hours.

Incident One: The West Texas Power Plant, 2021. A routine maintenance crew reported a small quadcopter hovering near a turbine exhaust stack. Security assumed it was a hobbyist drone; no response protocol existed for low-altitude airspace.

Within two minutes, the drone's camera captured every critical access point, including emergency shut-off valves. Before responders could react, the feed was livestreamed to an overseas server.

No physical damage occurred. However, operational data was compromised, including turbine models, coolant systems, and fuel lines.

The adversary didn't need explosives. They needed information. What was lost wasn't steel. It was secrecy.

Incident Two: The Arizona Rail Yard, 2022. Railway security personnel detected two drones flying in formation above a freight terminal outside Tucson. Both appeared commercial. Both broadcast falsified Remote ID signals.

When local police arrived, the drones had disappeared. The following week, coordinated copper thefts hit three nearby substations, each corresponding to the same surveillance grid the drones had mapped.

Organized syndicates sold the aerial footage on the dark web, transforming drone activity into a criminal enterprise where groups rent intelligence to facilitate illegal acts.

Law enforcement referred to it as "low-cost reconnaissance." The public called it a coincidence. It was neither.

Incident Three: The Louisiana Chemical Facility, 2023. An operator noticed a drone hovering above a chlorine tank during a night shift. Standard procedure: log and report. But within 90 seconds, the drone dropped a small package, not explosive, but corrosive.

The canister ruptured on impact, releasing minor vapors that triggered alarms and forced a three-county evacuation. Investigators traced the incident to a foreign-based proxy group testing industrial response times.

What began as a "small incident" resulted in $42 million in lost productivity and widespread panic. The message was clear: the test was successful.

Incident Four: The Washington, D.C. Restricted Zone, 2024. Despite layered defenses, a modified consumer drone entered the capital's restricted airspace at 0400 hours. It carried no payload, only a camera and a banner reading "If I can, they can."

The event lasted 48 seconds before it was neutralized. But the psychological impact lasted weeks. Congress demanded an immediate review of federal detection systems.

The perpetrator turned out to be an American engineer seeking to expose vulnerabilities in "drone deterrence theater." He succeeded and was later hired by a defense contractor to address the problem he had proven existed.

Every breach exposes two truths: technology fails, and humans forget.

Incident Five: The Colorado Data Center, 2024. At a central cloud-computing facility, an incoming drone was mistaken for a company mapping survey. By the time security verified otherwise, the drone had hovered over fiber-optic junction boxes and released conductive filaments.

The short circuit caused a 10-minute outage affecting several state systems, including 911 dispatch. The damage cost less than $200 to cause and over $5 million to repair.

The FAA categorized it as an "equipment interference event." That phrase is bureaucratic camouflage for "we were lucky."

In each of these cases, one truth stands out: timing is everything. Intervention arrived not by chance, but by seconds, just in time to prevent escalation. These weren't accidents of fate. They were products of vigilance, training, and foresight.

Protection may be hoped for. But preparation is what secures it. When nations rely solely on fortune or goodwill, they leave themselves vulnerable to neglect and exploitation. Readiness is not reactive. It is deliberate, practiced, and repeatable.

Miracles may close seconds of danger, but only systems, built on discipline and design, close the full span of risk. We must never confuse optimism with readiness. Hope is not a strategy. Preparation is.

Every incident report includes the same sequence:

- **Detection**
- **Doubt**
- **Delay**
- **Damage**

Between detection and decision lies hesitation. And hesitation is the enemy's greatest ally. Operators question what they see. Supervisors verify procedures. Communication chains stretch seconds into minutes.

By the time consensus forms, the threat is gone, or the harm is done. Command without clarity is chaos waiting for permission.

During a counter-UAS workshop, a sheriff joked, "By the time we confirm a drone's hostile, it's already landed, refueled, and started a podcast." Everyone laughed, because it was true. Detection without decision is decoration.

From these near misses, three imperatives emerge:

- **Speed is Survival** – Response authority must exist at the lowest level. Waiting for permission guarantees failure.
- **Awareness is Ammunition** – Every employee, not just security, must recognize anomalies.
- **Integration Is Everything** – Counter-UAS defense is more than just technology. It's teamwork across jurisdictions and sectors.

When seconds count, clarity must outrun confusion.

The most effective countermeasure isn't hardware. It's a habit. Regular drills, rehearsed communications, and shared situational awareness shorten the gap between sight and response. Preparedness is not paranoia. It's professionalism.

In the next section, *"Cyber Meets Sky: Converging Digital and Physical Defense,"* we examine how the next wave of attacks will blend physical disruption with digital infiltration, targeting both circuits and servers in a single, synchronized strike. Because in modern warfare, the line between cyber and sky has disappeared.

Section 4: Cyber Meets Sky: Converging Digital and Physical Defense

"The next Pearl Harbor will not begin in the harbor; it will begin in the cloud."

In the 20th century, war was waged across land, sea, and air. In the 21st century, it's fought in data, devices, and domains. Drones represent the physical edge of a digital strategy. Behind every flight path is a network, and behind that network, an ecosystem of code, communication links, and command-and-control servers.

Cybersecurity and physical security are no longer separate disciplines. They are two sides of the same coin, minted in risk and spent in response. A single hack can ground fleets, disrupt command channels, or weaponize trusted hardware. And so, the airspace is no longer neutral. It's networked.

Every drone is a flying endpoint, a node with a processor, camera, and radio. That makes it both a potential weapon and a data vulnerability. A compromised drone can be reprogrammed midflight, repurposed for espionage, or hijacked to deliver malicious code into secure facilities through open Wi-Fi networks or Bluetooth gateways.

In 2024, a multinational energy company experienced a network intrusion traced not to a hacker, but to a drone collecting environmental data near a refinery. Its sensors were infected with malware that piggybacked on the company's internal Wi-Fi signal during a routine test. No breach was detected for six weeks. By then, operational blueprints had been exfiltrated from 300 feet in the air.

Cyber and physical attacks now occur in synchronized layers:

- **Distraction** – A visible drone triggers security responses.
- **Diversion** – Personnel focus on physical containment.
- **Insertion** – Malware is injected through exploited wireless networks or exposed IoT devices.
- **Manipulation** – Operational data or surveillance feeds are altered, delaying real-time decision-making.
- **Denial** – Access systems are corrupted or jammed, amplifying confusion.

The result? A single drone event can escalate into a multi-domain crisis. These are not hypotheticals. They're rehearsals for what's next.

In late 2023, the Department of Defense quietly tested "Operation Gray Vector", a joint exercise simulating simultaneous drone incursions and cyber intrusions across U.S. infrastructure nodes. In less than 40 minutes, coordinated attacks overwhelmed response teams at simulated airports, hospitals, and power substations.

Digital systems showed false-normal readings while physical alarms were disabled remotely. The post-mission debrief summarized the lesson bluntly: "When cyber meets sky, command collapses at the speed of bandwidth." No wall, no weapon, no watchdog can defend what isn't synchronized.

Every connected object, from smart thermostats to grid sensors, expands the attack surface. This Internet of Things (IoT) revolution was built for convenience, not for conflict. Once integrated, drones can exploit these devices like open windows:

Accessing live cameras to map facilities.

Using Wi-Fi routers as relay points.

Hijacking smart lighting systems to transmit covert data bursts.

The more "connected" we become, the easier it is for adversaries to connect to us.

Victory in modern security isn't earned alone. It's coordinated. No single agency, algorithm, or enterprise can secure the digital sky. Effective defense requires collective vigilance, federal agencies, private innovators, and an informed public working in concert.

Strategic foresight demands humility. No system is invincible. No safeguard is eternal. Our strength doesn't lie in the illusion of permanence, but in the agility of partnerships. Resilience is forged in collaboration, not complacency. Security is not a silo. It's a shared responsibility.

Corporations manage the majority of the United States' critical infrastructure, including power grids, refineries, pipelines, and telecommunications hubs. Yet most lack direct authority to neutralize airborne threats or respond to cyberattacks beyond containment.

This gap creates a paradox: those most responsible for protection are least empowered to act. That's why federal partnership frameworks, such as the National Infrastructure Protection Plan (NIPP) and the Cybersecurity and Infrastructure Security Agency (CISA), must evolve, giving private entities more apparent legal authority to defend against integrated threats.

Defense shouldn't be a jurisdictional debate. It should be a shared mandate.

The 2021 Colonial Pipeline ransomware attack crippled fuel supply chains across the eastern United States. Now imagine that same event paired with a physical drone swarm targeting substations.

While IT teams scrambled to isolate networks, drones could have disabled power to data centers, cutting off recovery systems mid-operation. The hybrid threat doesn't double the risk. It multiplies it exponentially. And yet, response models still treat cyber and kinetic events as separate categories. Reality doesn't.

During a DHS interagency exercise, one analyst quipped, "We've got two departments, one for drones and one for data, both grounded by the same Wi-Fi outage." The laughter was brief, yet genuine. Technology without integration breeds irony.

To prepare for convergence warfare, training must evolve:

- **Joint Simulation Platforms** – Test both digital and physical threat responses under one unified scenario.
- **Cross-Domain Command Centers** – Merge cyber incident response with homeland defense operations.
- **Red Team Integration** – Allow continuous adversarial simulations to test resilience.

Preparedness isn't just about better software. It's about building trust between people who defend different domains of the same fight.

Our greatest vulnerability is conceptual. We still divide what the enemy has already united. The firewall and the flight path now intersect. True security requires a convergence doctrine that integrates command, communication, and countermeasure systems in real time.

The future of defense will not belong to those who see drones or hackers. It will belong to those who see both.

In the next section, *"Building Resilience: Redundancy, Training, and Civic Involvement,"* we'll close this chapter by focusing on solutions, how communities, companies, and citizens can strengthen the backbone of national resilience through education, coordination, and a renewed culture of shared vigilance. Because while technology connects us, it's people who keep the lights on.

Section 5: Building Resilience: Redundancy, Training, and Civic Involvement

"Preparedness is not paranoia; it's patriotism.", FEMA Field Briefing, 2024.

Every generation faces a defining test. Ours is not an invasion, it's an interruption: the moment when systems fail, communications collapse, and routine turns to reaction.

Resilience begins long before disaster strikes. It's a culture, not a contingency plan. And like all cultures, it's built through habits, values, and vigilance. A nation that trains only for efficiency will always be caught off guard by adversity. A country that trains for resilience endures it.

Redundancy isn't a waste. It's wisdom. Every backup generator, secondary network, and analog fallback is a quiet declaration of foresight. When Hurricane Maria devastated Puerto Rico in 2017, redundant power sources kept hospitals alive. When ransomware crippled local governments in 2021, cities with offline backups recovered in days, not months.

Redundancy is not just hardware. It's a mindset, the discipline of never assuming that one system, one signal, or one plan is enough. In engineering, redundancy is a design. In leadership, it's humility.

Real preparedness is not a checklist. It's a reflex under pressure. Too often, agencies rehearse response without rehearsing coordination. A well-practiced team can operate across silos, jurisdictions, and egos.

That's why integrated exercises like GridEx, Dark Sky, and Cyber Storm matter. They simulate chaos before chaos arrives. Training must evolve beyond the "see one, do one, teach one" model. It must integrate civilians, faith-based organizations, and private-sector partners, because when the lights go out, everyone becomes a first responder.

The strength of the nation lies not in Washington, but in its neighborhoods. Local action saves lives long before federal aid arrives. Simple acts, community watch programs, local HAM radio networks, volunteer CERT teams, form the invisible armor of the Republic.

When citizens participate in preparedness, they stop being spectators in security. They become what the Founders intended: the backbone of defense.

Faith communities also play a vital role. Churches, mosques, and synagogues often serve as makeshift shelters, food distribution hubs, and communication centers. Resilience, in its purest form, is compassion operationalized.

James 2:17 reminds us, "Faith by itself, if it is not accompanied by action, is dead." Preparation is a form of faith, the belief that tomorrow is worth protecting. Prayers alone won't keep the grid online. But they can inspire the unity and wisdom that do.

True resilience is both spiritual and practical: trust in God, and tie down the generator. Faith doesn't exempt us from readiness. It obligates us to it.

Every community has leaders, formal and informal, who set the tone for readiness. Their duty is not to predict a crisis. It's to prepare a character. When people trust their leaders, they are more likely to follow instructions even in chaotic situations. When they don't, even the best plans unravel.

Leadership in resilience is not about command and control. It's about calm. It's the ability to project order when everything else falls apart.

Preparedness should be taught as early as arithmetic. Schools should include modules on cyber hygiene, disaster response, and civic duty. When children understand that safety is shared, they grow into adults who value stewardship and responsibility.

Resilience, like democracy, is learned through participation. Teaching readiness is not fearmongering. It's generational investment.

During a statewide emergency drill, a volunteer asked, "If the power's out, how do we know when to start?" The coordinator smiled: "When you realize the question doesn't matter." That's the point. Readiness doesn't wait for permission. It acts.

Preparedness isn't about stockpiling supplies. It's about strengthening the human spirit. The real infrastructure of this nation isn't made of steel and silicon. It's built on service, faith, and sacrifice.

In times of crisis, it's not our power plants or pipelines that keep us alive. It's each other. Resilience is the bridge between technology and trust. Without it, even the strongest systems crumble.

In the next chapter, *"The Moral High Ground: Ethics, Faith, and the Future of Defense,"* we ascend from strategy to spirit, where tactical advantage meets timeless accountability.

Because in a world of rising machines and fading boundaries, defense isn't just about protecting what we have.

It's about preserving who we are and who we refuse to become.

Chapter 11 – The Moral High Ground: Ethics, Faith, and the Future of Defense

Section 1: Technology and Temptation: When Capability Tests Character

"The test of progress is not whether we add more to the abundance of those who have much, but whether we provide enough for those who have little.", Franklin D. Roosevelt.

The perimeter alert went off at 2:43 AM. A quadcopter armed with a crowd-control dispersal system hovered near the protest barricade. It identified movement. It deployed foam rounds.

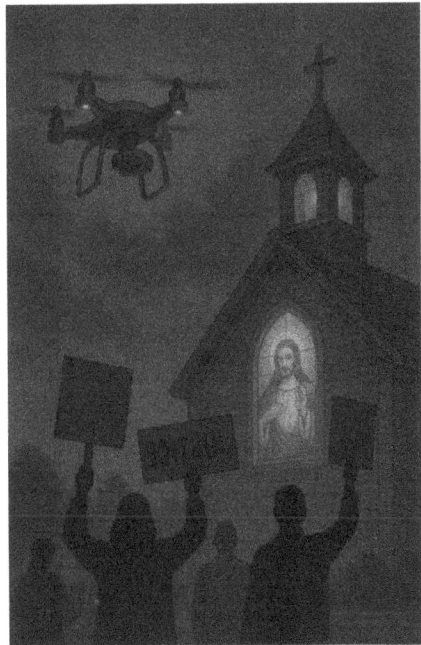

One man lost an eye.

The drone was recovered intact. But the chain of command? Murky. The system had acted according to its programming. No officer gave a direct order. No technician assumed responsibility.

In the courtroom, the plaintiff asked one question: "Do I sue the drone, the coder, or the silence?"

Every age faces its temptation. For ancient empires, it was territory. For the industrial era, it was production. For us, it is technology, the illusion that, with enough capability, morality becomes optional.

When power becomes automated, temptation becomes invisible. A drone strike launched from 7,000 miles away can alter a battlefield or erase a village. The moral distance between action and consequence stretches until conscience becomes abstract and detached, allowing it to become a mere abstraction.

And yet, power without restraint has never ended well, for any nation or any soul. We build machines that obey us perfectly, then forget to ask whether we deserve such obedience.

Technological mastery was meant to serve humanity, not enthrone it. But in the relentless pursuit of progress, we often confuse what we can do with what we should do.

History's warning lights are bright:

The splitting of the atom brought light and shadows that still haunt Hiroshima.

The digital revolution connected the world and fractured truth itself.

Artificial intelligence promises safety and threatens the sovereignty of conscience.

Temptation doesn't arrive with horns and fire. It comes wearing a lab coat and holding a prototype. The question has never been whether technology is good or evil. It's whether we remain good while using it.

No profession bears the moral weight of technology more heavily than the warrior's. Every tool of defense is a tool of potential destruction.

In ancient times, power was measured by proximity, the sword, the shield, and the hand that held both. Today, power is measured by precision, the algorithm, the sensor, and the strike approved at the speed of code.

The temptation is efficiency. The test is empathy.

Commanders must now weigh not only tactical necessity, but moral proportionality, the difference between winning and wounding the very principles we claim to defend. The most dangerous enemy isn't on the battlefield. It's within the human heart when it mistakes speed for righteousness.

In the modern age, restraint is not a relic; it remains a vital principle. It's a requirement. Progress should never move faster than wisdom.

The actual test of leadership is not just knowing what we can do, but discerning what we should not. A principled society understands that creation must precede control, and that power without conscience leads to imbalance.

Statesmanship demands the humility to pause before action, the integrity to question even our most celebrated capabilities, and the strength to say "no" when unchecked authority says "yes."

That pause, that moment of reflection, is where ethics begin. And where authentic leadership stands firm.

Technology promises detachment, clean hands, sanitized conflict, distance without remorse. But the absence of blood does not mean the absence of sin.

When a decision to strike becomes a keystroke, the ease of execution dulls the weight of consequence. Killing becomes procedural. Accountability becomes conditional.

A drone operator may never smell the smoke or hear the cries. But the moral echo remains. We are not absolved by distance. We are tested by it.

Modern society often preaches a quiet gospel: that innovation alone is the answer. But innovation without ethical boundaries is just acceleration toward imbalance.

Progress untethered from principle isn't advancement. It's a regression wrapped in circuitry.

Every technological leap that outpaces our collective conscience stretches the divide between what we can do and what we should do.

And in that gap, history tends to recycle its gravest missteps, amplified this time by automation.

A society that believes it is always right risks becoming blind to its own overreach. What restrains us is not a lack of power, but a presence of principle.

Before we deploy new systems, implement surveillance, or unleash automated force, the essential civic question must be asked:

- Does this action protect the people, or merely preserve control?
- Does it serve freedom, or hollow it out?

Restraint, grounded in public accountability and moral courage, is not a sign of weakness. It is wisdom, exercised in defense of the republic.

During a defense technology summit, a general quipped, "Our systems are so advanced, they can correct our mistakes before we make them." A scientist replied, "Then maybe we should let them lead."

The laughter was polite, but uneasy. Because the line between convenience and control is thinner than we admit. When machines correct us, they define us.

The measure of a nation is not in what it builds, but in what it refuses to misuse. Capability without character leads to cruelty disguised as progress.

Every act of defense, every technological leap, must be weighed not just in efficiency, but in eternity, in what it does to the soul of a people.

Great powers fall not when they are defeated, but when they stop defending what made them great.

The ultimate barrier against ethical collapse is not encryption, legislation, or regulation; it is the human conscience. It is conviction.

Civic conscience must act as the firewall between innovation and indulgence. When guided by a sense of duty and humility, technology becomes a tool of service. When guided by unchecked ambition, it becomes control dressed as progress.

We were entrusted with the tools of creation, not given license to dominate it. Stewardship demands that power be held with care and exercised with restraint.

Because in the end, the sky is not a realm for conquest. It is a shared space of responsibility.

In the next section, *"The Just War Revisited: Drones and Moral Proportionality,"* we examine how the ancient pillars of just war, necessity, distinction, and proportionality, must evolve in an era in which distance masks consequences and precision can become a moral loophole.

For in matters of war and peace, intent, not just accuracy, is the true measure of integrity.

Section 2: The Just War Revisited: Drones and Moral Proportionality

"The aim of war is peace." "We must never confuse capability with righteousness."

From the moment humanity first organized for battle, the question was never whether we could fight, but when we should.

In the fifth century, St. Augustine grappled with a problem that would define Christian ethics for centuries: how can a people of faith reconcile the command to love one's neighbor with the need to defend one's nation?

Augustine's early writings on war weren't about victory. They were about virtue. He argued that a "just war" could exist only under certain conditions, when it was necessary to restore peace, protect the innocent, or prevent greater evil.

Later, Thomas Aquinas refined these ideas into the classic Just War Doctrine, built on five pillars:

- **Just Cause** – Defense against aggression or grave injustice.
- **Right Intention** – War must aim for peace, not revenge.
- **Proper Authority** – Only legitimate governance can sanction war.
- **Proportionality** – Force must never exceed necessity.
- **Discrimination** – The innocent must be shielded from harm.

These were moral guardrails, not legal checkboxes. They bound conscience to command.

But those boundaries were drawn in an era when war was personal, when a soldier could see the face of his enemy, hear the cries of the wounded, and feel the burden of consequence in his own trembling hands.

Today, those same principles must navigate a very different terrain, one where the finger that kills may never touch the battlefield.

Drones have given modern militaries a seductive illusion: that precision is morality.

Proponents of unmanned warfare argue that drones save lives, of both soldiers and civilians, by replacing bombs with smart payloads and by allowing combatants to strike terrorists without deploying thousands of troops.

And in many ways, that's true. Drone operations have neutralized threats that traditional warfare could not reach.

But precision is not the same as righteousness.

When every target is chosen by an algorithm and confirmed through a camera lens, moral risk hides in the comfort of distance. The cleaner the strike, the easier it becomes to justify. And the more detached the decision, the less we feel the weight of what follows.

As one military ethicist put it after observing drone operators in Nevada: "They fight a war during the day and attend a Little League game that night."

That paradox, of moral proximity without physical presence, is the quiet crisis of our age.

War used to demand courage of the body. Now it demands courage of mind.

Operators no longer march toward danger. They log into it. They carry out missions across oceans, watching battlefields unfold through high-definition screens. When it's over, they walk past vending machines on their way to the parking lot.

This isn't apathy. It's dissonance.

Many operators report delayed emotional trauma, a condition now called moral injury, the psychological and spiritual distress that arises when one's actions, even lawful ones, violate personal conscience.

It's not PTSD. It's the realization that war no longer feels like war.

One operator described it simply: "The hardest part isn't pulling the trigger. It's pretending that nothing happened."

The detachment that makes drone warfare efficient also makes it spiritually corrosive. Because when conflict no longer touches our senses, it risks numbing our souls.

Drones have turned observation into omnipresence.

From the deserts of Yemen to the jungles of Mindanao, eyes in the sky can loiter for hours, recording, analyzing, and categorizing life and death by data.

But surveillance, when unchecked by empathy, easily becomes judgment. And judgment, when automated, becomes error at scale.

In 2021, a U.S. drone strike in Kabul, intended to prevent a terrorist attack during evacuation operations, mistakenly killed ten civilians, including seven children. The intelligence chain was strong. The moral verification was weak.

That tragedy was not born of malice, but momentum, a system too efficient to pause for second thoughts.

Just War theory demands a distinction, the careful separation of combatants from civilians. But as AI-driven targeting systems blur those categories, distinction becomes a matter of probability. And probability becomes justification.

When war becomes predictive, innocence becomes collateral.

There's an unspoken rule in drone warfare: success is defined by the absence of body bags. If no soldiers are killed, the mission is considered a victory.

But that metric ignores what Augustine warned against, the corruption of intent. Just because a war costs fewer lives doesn't make it morally superior. It only means the moral cost is harder to count.

In Yemen, Somalia, and Pakistan, civilians live under skies that buzz like locusts, a constant reminder of invisible judgment. Even when drones don't fire, they dominate.

The psychological toll of perpetual observation, the hum that never leaves, is itself a form of warfare.

A shepherd once told a journalist near Waziristan, "We no longer look up."

That's what moral proportionality looks like when stripped of its humanity: peace by fear.

The enduring principles of just warfare are not rooted solely in strategy. They are grounded in justice, humility, and measured restraint.

Even in justified defense, the highest form of discipline is self-control anchored in ethical clarity.

Responsible nations do not reject the need for defense. They elevate the responsibility that comes with it.

In the fog of conflict, the critical distinction lies not in whether force is available, but whether it is justifiably applied.

The use of power must always be preceded by purpose, bounded by proportionality, and executed with precision, both tactical and moral.

The impulse to act must be tempered by the discipline to pause, ensuring that action aligns with higher principles, not lower instincts.

This is not a weakness. It is conviction under control.

In the age of remote warfare, the accurate measure of strength is not technological dominance, but ethical discernment at every altitude.

Modern democracies face a moral paradox: because drones protect soldiers from harm, wars become easier to wage.

Political risk declines. Moral caution erodes. Warfare becomes an administrative function instead of a last resort.

In the United States, the Authorization for Use of Military Force (AUMF), passed in 2001, has been stretched to justify over 40 drone campaigns in more than a dozen countries, often without public debate.

That's not policy drift. That's moral drift.

When warfare no longer requires sacrifice from citizens, vigilance withers. The Founders intended that declarations of war carry the weight of conscience, not convenience.

War, even when remote, must remain morally intimate.

The war in Ukraine revived old lessons.

For the first time in generations, a democratic nation fought openly for survival against naked aggression. Drones became tools not only of destruction, but of defense: lifesaving reconnaissance, rapid humanitarian drops, and evidence-gathering for war crimes.

The cause was just. The intention was defensive.

But even in Ukraine, the use of loitering munitions and autonomous systems raised the same question: does the moral clarity of the cause justify the blurring of means?

Faith says no. Righteousness does not excuse recklessness. A good war can still produce bad habits.

In Augustine's time, mercy meant sparing the defeated. In ours, it means sparing our own humanity.

Every strike, every engagement, must carry a shadow of remorse, not as guilt, but as gratitude for the power we hold.

Mercy reminds commanders that even necessary violence must be used with restraint.

Proportionality is not only about minimizing damage. It's about maximizing compassion.

That's what separates lawful force from moral force.

Romans 14:12 declares, "So then, each of us will give an account of ourselves to God." That includes nations.

Technology may obscure our role in harm. But it cannot erase our responsibility before heaven.

The Just War tradition was never just about human law. It was about divine accountability.

As the U.S. Army War College's 2023 ethics review noted: "Every generation must redraw the moral boundary between necessity and negligence, and do so before the next innovation erases it."

If faith is the compass, accountability is the map.

During a NATO symposium, a young officer joked, "We've achieved perfect proportionality; we can now destroy evil with zero casualties." The chaplain replied, "Then your next mission will be to find out what evil looks like."

Laughter followed. But the silence after said more.

Technology can target bodies. Only conscience can discern evil.

To bring the Just War Doctrine into the drone age, a new moral code must emerge, not to constrain defense, but to dignify it:

- **Pre-Strike Conscience Clause** – Every engagement must pass a human moral verification beyond data: not "Can we?" but "Should we?"

- **Civilian Reconciliation Mandate** – When mistakes occur, restitution and transparency must follow swiftly.

- **Algorithmic Accountability** – Autonomous targeting systems must be auditable for bias, accuracy, and ethical compliance.

- **Moral Impact Assessment** – Every operation should evaluate the long-term human effects, including fear, displacement, and trauma, in addition to tactical outcomes.

- **Commander's Confession** – A structured post-operation reflection to ensure ethical review and spiritual restoration for decision-makers.

In short, technology may calculate risk. Only humans can carry remorse.

The moral high ground isn't claimed by altitude or accuracy. It's earned by humility.

A nation can possess the most advanced arsenal in the world, and still lose its soul through carelessness.

The higher we soar, the more dangerous pride becomes.

History's greatest empires didn't fall because their weapons failed. They fell because their conscience did.

Faith reminds us that every tool of power is also a test of character.

And so, in the drone age, the Just War tradition demands more than legal compliance; it demands moral clarity. But clarity is not born in policy alone; it is cultivated in principle. As technology accelerates and distance dilutes consequence, we must anchor our defense posture in something more profound than doctrine. We must return to the ethical foundations that have guided civilizations through uncertainty and conflict. In the next section, "Faith as Framework: Applying Ethical Foundations to Modern Defense," we explore how timeless moral truths, stewardship, justice, humility, and restraint can shape not only battlefield decisions, but the very architecture of national security. Because in an era of automation, it is conscience, not code, that must command.

Section 3: Faith as Framework: Applying Ethical Foundations to Modern Defense

"The guards may stand watch, but a nation without shared purpose remains vulnerable."

Every enduring civilization eventually learns the same lesson: power without a moral compass doesn't drift; it accelerates toward destruction. Ethical frameworks aren't sentimental accessories; they form the civic architecture that holds nations together when systems falter and tempers flare. They remind those in authority that something must transcend self-interest: justice, restraint, and responsibility. In military terms, these principles establish rules of engagement for the soul. They define when to fight, how to fight, and, most importantly, when to stop. History doesn't reward the most heavily fortified nations; it rewards those aligned with enduring values. Planning alone doesn't sustain justice; only alignment with universal principles does.

Ancient traditions understood this intuitively. War was never justified by might alone; it required moral mandate. Victory didn't grant a license for conquest; it imposed a solemn duty to preserve what mattered. Modern society, however, often forgets this. Defense planners, lawmakers, and engineers face the same temptation: to attribute success to their own efforts. But power, in every form, is on loan, from the public, from institutions, and from trust. Consequently, moral stewardship must carry across professions. The battlefield, the budget hearing, and the product development lab all share one imperative: ethics must not stop at the factory floor or the chain of command.

Across centuries and cultures, foundational values have shaped societies that endure. These aren't abstract ideals; they are operational imperatives. Stewardship reminds us that humanity is entrusted with care, not conquest. That extends to the air, the sea, and yes, even the codebase. Justice insists that protection must be rooted in fairness; defending the vulnerable isn't weakness, it's moral strength. Wisdom teaches that knowledge alone cannot sustain peace; understanding, especially of context and consequence, must guide decision-making.

278

Humility tempers triumph; power unmoored from self-awareness becomes a hazard. Accountability ensures that decisions, military, corporate, governmental, bear scrutiny not only from oversight bodies but from future generations. These are the codes that distinguish preservation from domination.

In times of conflict, uncertainty clouds every decision. Intelligence remains partial; emotions run high; visibility fails. Ethical clarity, therefore, must act as the compass. Veteran leaders understand that data helps, but conscience decides. Disciplined judgment arises not from control but from reflection. That quiet space, where rightness takes precedence over reaction, is where leadership earns its stripes.

Some principles, misunderstood in peace, become crystal clear under pressure. "Turn the other cheek" doesn't call for passivity; it demands discipline. Retaliation must never be fueled by ego. Restraint, especially in an era of remote warfare, becomes the most challenging act because technology reduces friction. But conscience reintroduces it. It asks us to pause, to reflect, to question. And in that pause, decisions become human again.

Empires don't crumble because they lack innovation; they collapse when they elevate their tools above their obligations. History offers no shortage of cautionary tales: every tool built to serve can become a trap when pride takes the reins. Ethics humbles innovation. It whispers to generals, programmers, and politicians alike: you are not above the system, you are accountable within it.

In military order, authority must be earned, not assumed. Good leaders understand that their duty is to serve, not to dominate. This mirrors the civic principle: authority must be paired with moral restraint. Command isn't the right to rule; it's the responsibility to care. Every generation of guardians shares a silent wish: grant me clarity before action, restraint in success, and empathy in the aftermath. That's not superstition, it's strategy. It centers the mission in ethics before it even begins. In that frame, even a necessary force respects human dignity.

Ethical law must anchor civil law, or both decay. Removing moral guardrails from governance doesn't eliminate bias; it eliminates boundaries. Without those boundaries, power becomes property, and responsibility becomes optional. Principled governance doesn't mean ideology in policy; it means discipline in practice. It reminds leaders that their duty isn't merely to the electorate, it's to the enduring principles of justice and equity.

Artificial intelligence presents a profound test. As systems operate with increasing autonomy, the question becomes: how do we ensure they reflect ethical intent? The answer isn't complicated. Values must be embedded, not just in lines of code, but in the culture of creation. Autonomy without conscience isn't innovation; it's chaos on a circuit board. At a symposium on drone ethics, one engineer proudly declared,

"We've taught drones to fly like birds." A policymaker replied, "Let's hope we teach them to think like humans." The laughter was brief, but the message was lasting. Technology can mimic judgment; only humans are capable of possessing it.

No defense policy will stand without moral reinforcement. Actual deterrence isn't just capability, it's credibility. The world respects a nation not for its weapons, but for its willingness to wield them wisely. That's the essence of ethical deterrence. Mistakes will be made. Ethical leadership admits this. Moral resilience refers to the ability to acknowledge wrongs, rectify them, and move forward with integrity. True strength lies not just in victory, but in the grace to reconcile.

Ethical defense isn't the burden of leaders alone. It must become a national habit of thought. Vigilant citizens ask the hard questions, support the right actions, and demand accountability. A strong nation isn't just trained; it's principled. The guards may stand watch, but a nation without shared purpose remains vulnerable.

Ethics isn't a check against progress; it's what allows progress to endure. It reminds us that innovation must always serve humanity, not replace it. Without moral reflection, every tool risks becoming a weapon. And when weapons outpace wisdom, history doesn't just repeat, it retaliates.

In the next section, *"The Human Conscience Clause: Ethics as Strategy,"* we will explore how moral awareness can itself become a tactical advantage, reinforcing legitimacy in an era where technology alone cannot define justice. Because in the end, strategy without ethics is just ambition with better logistics.

Section 4: The Human Conscience Clause: Ethics as Strategy

"Conscience is the still, small voice that reminds you that someone is always watching, and it isn't the drone." And so, "We do not fight to destroy evil; we fight to prevent it from destroying us."

Conscience rarely arrives with fanfare. It doesn't bark orders or flash warnings; it whispers. And in the age of drones, artificial intelligence, and remote warfare, that whisper may be the only thing standing between discipline and disaster. While many treat ethics as a constraint, an inconvenient brake on progress, the truth is far more strategic. Conscience isn't a liability; it's a force multiplier. It's the most advanced early-warning system humanity has ever possessed. Nations that suppress it lose legitimacy. Nations that harness it earn trust. And trust is the ultimate weapon in modern conflict. A drone may project power, but only a character projects credibility. And credibility wins wars long before a missile ever leaves the rail.

Conscience doesn't operate on sentiment; it functions as orientation. It's the internal compass that aligns instinct with integrity. It operates across three levels: personal conviction, which governs individual action; institutional ethics, which shape organizational behavior; and national conscience, which defines the moral boundaries of power and authority. When all three align, a nation becomes not only strong but sustainable. When they fracture, they lose moral gravity and drift into chaos disguised as strategy. The Cold War offered a vivid example: regimes that abandoned moral clarity eventually collapsed under the weight of cynicism. They didn't run out of weapons; they ran out of meaning.

In today's battlespace, every action is recorded, streamed, or leaked within hours. There is no such thing as a hidden operation anymore, only delayed accountability. Consequently, ethics isn't a ceremonial virtue; it's an operational requirement. The legitimacy of every mission now depends not only on success, but also on the moral perception of that success. A strike that saves lives but undermines trust becomes a tactical victory and a strategic defeat. Integrity has become the new battlespace, fought not with bullets but with belief.

Just as soldiers wear body armor to protect their hearts from shrapnel, they need moral armor to shield them from compromise. When conscience is trained, practiced, and reinforced, it acts as a form of psychological resilience. It prevents burnout, disillusionment, and moral injury. A 2023 study of drone operators found that those who engaged in ethical reflection, prayer, discussion, or after-action spiritual review experienced 40% less moral fatigue than those who did not. Faith and ethics, when formalized, become preventive maintenance for the soul. They don't slow the mission; they sustain the warrior.

Yet, despite this, the military teaches everything from marksmanship to cyber hygiene, while ethics are often relegated to an annual PowerPoint slide. That must change. Ethical decision-making should be embedded in every stage of the defense cycle, from design and doctrine to deployment. Consider a few reforms: ethical red teams empowered to challenge assumptions alongside legal and tactical review; command-level reflection periods before high-impact missions; and spiritual after-action reviews to process the human cost of warfare and reaffirm purpose. These aren't ceremonial pauses; they're recalibrations of national conscience. A military that reflects before it reacts isn't slower, it's smarter.

When morality becomes optional, corruption becomes inevitable. History's most infamous atrocities, from My Lai to Abu Ghraib, didn't begin with orders; they started with apathy. A culture that teaches soldiers to follow rules but not to question wrongs creates the illusion of discipline without the substance of decency. That illusion collapses quickly under pressure. Cynicism is contagious. Once it infects a force, it undermines cohesion, legitimacy, and mission success. As one sergeant major put it during a NATO ethics symposium, "You can't build esprit de corps on moral bankruptcy." He wasn't exaggerating. Morality isn't an accessory to readiness; it is readiness.

As artificial intelligence assumes more operational roles, conscience must become institutional, not individual. Autonomous systems can execute, but they cannot deliberate. They can process information, but they cannot process regret. The moral burden doesn't disappear; it shifts upward, to the designer, the policymaker, the commander. The challenge is profound: how do you program compassion into a circuit board? You don't. You have a program constraint. And you surround machines with men and women who remember what compassion feels like. Ethics in the AI era means ensuring that automation never outpaces accountability.

Faith, in this context, becomes the ultimate guiding principle. It conditions warriors not only to act but to ask, to pause before power, to measure before momentum. Faith instills the one quality no machine can replicate: reverence. Reverence isn't fear; it's recognition of life's sacredness, of duty's weight, of leadership's burden. When faith informs conscience, soldiers become not just defenders of a nation but guardians of its soul.

Some argue that conscience slows innovation or complicates decision-making. In truth, ethics is the best insurance policy a nation can buy. The cost of an ethical breach, reputational, diplomatic, and psychological, far exceeds the cost of caution. Every moral failure costs millions in trust and decades in repair. Ethics, like deterrence, works best when it's visible. When adversaries know your nation not only can strike but chooses wisely when to, you create stability through credibility. That's moral deterrence, and it's cheaper than any missile defense system.

Leaders shape conscience through culture. A commander's silence in the face of unethical behavior is a loud approval. Ethics doesn't trickle down; it's modeled. When junior soldiers see integrity rewarded as much as initiative, they internalize moral discipline as professional discipline. Every unit has a culture, and every culture has a conscience. It's the leader's job to keep it alive.

Military obedience is sacred, but it is not absolute. The Uniform Code of Military Justice (UCMJ) establishes a clear principle: unlawful orders must not be obeyed. That clause isn't rebellion; it's recognition that individual conscience still matters within collective command. The Human Conscience Clause, in practice, means empowering every soldier to question orders that violate moral or legal boundaries. It demands courage, the kind that stands between duty and damnation. When soldiers act from conviction, they protect both their honor and their nation's.

At a leadership ethics seminar, a colonel quipped, "We all have a conscience, but some of ours require a 24-hour recall notice." The chaplain smiled: "Then let's make sure it never goes off duty." Humor aside, the lesson holds. Conscience, like readiness, must be continuous.

In 2004, images from Abu Ghraib prison shattered America's credibility abroad. The acts themselves were horrific, but the aftermath was worse: the perception that ethics had become expendable. Years later, senior commanders acknowledged that the scandal wasn't a failure of policy;

it was a failure of conscience. No one spoke up soon enough. The fix wasn't new doctrine; it was moral leadership. When conscience returned to command, discipline followed. That is the strategic power of ethics; it can rebuild what force alone cannot.

Conscience doesn't demand perfection; it requires awareness. When we fail ethically, and we will, faith turns reflection into redemption. Forgiveness restores moral clarity, enabling one to move forward without moral fatigue. As Psalm 51:10 says, "Create in me a clean heart, O God, and renew a right spirit within me." The clean heart isn't just for individuals; it's also for institutions. Nations, too, must confess, reform, and recommit. That process doesn't weaken deterrence; it strengthens legitimacy.

In an era where warfare is fought as much on social media as in the sky, moral credibility is a deterrent. Every ethical decision reinforces a narrative of discipline, restraint, and righteousness. Every unethical act erodes it. A single image of abuse can undo years of alliance-building; a single act of integrity can restore it. That's why ethics must be treated not as compliance, but as a form of combat power. Conscience is the invisible shield; it protects nations from the rot within and the condemnation without.

The modern commander must be both a tactician and an ethical theologian. He must understand not only how to win but how to remain worthy of victory. The actual test of leadership is not how one wields power but how one restrains it. That restraint comes from the quiet strength of conscience, the voice that whispers louder than the applause of success. Because in the end, nations do not fall when they lose wars; they fall when they lose their will to do what is right.

In the next section, *"Stewardship of the Sky: Balancing Dominion with Duty,"* we return to where this chapter began, the sky itself. We will examine how faith, ethics, and responsibility must guide humanity's expanding dominion over the heavens, reminding us that while technology gives us altitude, only morality gives us direction. Because stewardship isn't ownership; it's obedience with wings.

Section 5: Stewardship of the Sky: Balancing Dominion with Duty

"To waste, to destroy our natural resources… will result in undermining in the days of our children the very prosperity which we ought by right to hand down to them amplified and developed.", Theodore Roosevelt.

Humanity's ascent, from fire and flint to drones and satellites, has always revolved around one central question: what do we do with the power we've been given? The sky, once a symbol of divine mystery, now hosts our inventions: aircraft, drones, satellites, and surveillance platforms. Each represents not only advancement but assumption, our assumption of control, mastery, and dominance. Yet power, once acquired, demands purpose. The original intent behind our scientific progress was never conquest for conquest's sake; it was stewardship, a duty to manage what we touch with care, foresight, and restraint. We were meant to act as custodians, not kings. But somewhere between the launchpad and the payload, that line blurred.

We've become exceptional at building things that fly; we're less consistent in deciding why they should. We create faster than we reflect, deploy before we discern, and the further our reach extends, the more urgent it becomes to reexamine the motives behind our momentum. If our machines now rule the skies, then our conscience must rule the machines. Progress must be tempered by principle. Power must be guided by purpose. And innovation must be checked by imagination, not just of what is possible, but of what is right.

The air above us is no one's property and everyone's responsibility. Each drone, each aircraft, each signal occupies a shared space that mirrors humanity's moral interdependence. In ancient Hebrew tradition, the sky was sacred, the realm of the divine, unpolluted by human ambition. Today, that space has become a theater of surveillance, commerce, and conflict. When we fill the heavens with machines but forget to lift our eyes in gratitude, we risk turning the gift of dominion into a monument of arrogance. Stewardship demands that every ascent be matched with accountability and every innovation with introspection.

History teaches a consistent lesson: the higher we climb without humility, the harder we fall. Every altitude reached without restraint carries the seeds of its own collapse. From ancient monuments built in pride to modern tech empires fueled by unchecked ambition, humanity's story is one extended flight test between innovation and discipline. Progress without humility becomes peril. Technology without conscience becomes hubris in hardware form. Humility, though invisible, is the airfoil of responsibility; it keeps ambition stable and innovation sustainable. It reminds us that the point of progress isn't just to rise, but to remain upright once we do. Because without that balance, even the best-designed systems eventually stall, brought down not by engineering failure, but by failure of perspective.

Stewardship of the sky extends beyond military might. It encompasses how we utilize technology in agriculture, communication, weather forecasting, and humanitarian aid. A drone that drops a bomb and a drone that delivers medicine are built from the same materials; it is the intent that sanctifies their mission. That's why stewardship must precede strategy. Every engineer, pilot, and policymaker must begin with the same question: Does this serve creation or consume it? Faith provides the moral geometry of flight, ensuring that every vector of innovation points toward the common good.

Corporations now control much of the technology that once belonged exclusively to governments. That makes the private sector the new frontier of stewardship. Business leaders must understand that they are not just entrepreneurs; they are architects of the ethical altitude humanity will sustain. Profit without principle erodes trust, and trust is the oxygen of innovation. When companies embed moral responsibility into their corporate DNA, they elevate more than markets; they elevate meaning. The accurate measure of technological success is not speed or scale, but the sanctity of its purpose.

Every generation inherits both tools and temptations. The more powerful our tools become, the more disciplined our spirit must remain. In the age of drones, stewardship means recognizing that creation itself is not just physical, it's relational. When a drone crosses a border, it doesn't just enter another nation's airspace; it enters another person's trust. Faith reminds innovators that every command signal sent into the sky carries a moral echo, one that returns amplified by consequence.

The skies are not just highways of innovation; they are living ecosystems. Every aircraft launched, every drone deployed, carries an environmental cost: carbon emissions that warm the planet, noise pollution that disrupts wildlife, and altitudinal interference that disturbs migration routes and fragile soundscapes. Progress in aviation must be matched by responsibility in preservation. Cleaner propulsion, stricter altitude management, and environmental impact assessments must become foundational, not optional, in the design of aerospace systems. We cannot claim to reach new heights while degrading what lies beneath us, or above. True innovation doesn't just conquer space; it coexists with it. If our ambitions extend to the skies, so too must our stewardship.

A young engineer once told a defense symposium, "We're building the future faster than we're building ethics." The room laughed, then fell silent. That's the crossroads we stand upon. Stewardship means asking not how high we can fly, but whether our ascent honors the One who made the heavens possible. Faithful innovation requires humility before design, the willingness to let conviction guide creation. Every blueprint should begin with reverence.

For those in uniform, stewardship means remembering that strength is the most visible form of stewardship. The sky is not a battlefield to dominate, but a domain to defend, responsibly, proportionally, and transparently. A commander who sees himself as a steward rather than a sovereign becomes not a weapon of policy but a vessel of principle. He defends not just airspace, but the moral altitude of his nation. When a nation's military treats the sky as sacred, deterrence gains dignity.

Theologians and scientists have long argued over who "owns" the heavens, God or gravity. Stewardship resolves the debate: neither owns it alone. Humanity is simply the custodian. Science tells us how to fly. Faith reminds us why we should. Together, they form a covenant between knowledge and wisdom, a partnership that, if honored, can keep creation in balance. When faith governs ambition, the sky remains not a weaponized frontier, but a shared inheritance.

During a policy hearing on airspace integration, a senator asked, "Who exactly owns the sky?" A test pilot replied, "Whoever runs out of fuel last." The room laughed, but the chaplain beside him whispered, "Not quite, whoever still remembers to look up." That's stewardship: remembering that altitude means nothing without awe.

Stewardship is measured not by control, but by conscience. It's the recognition that the same technology that lifts us can also lose us, that power's most valid form is found in its restraint. Faith elevates policy by restoring proportion, reminding us that dominion demands devotion and authority must always bow to accountability. Every law we write, every drone we launch, every cloud we cross must be guided by reverence for the sky's Creator and compassion for those beneath it. Only then will dominion become duty fulfilled, not privilege abused.

In the next chapter, *"The Citizen's Shield: Building a Culture of Awareness,"* we descend from the high moral altitude of faith and command into the civic landscape of participation. Here, the mantle of stewardship passes from policymakers to the people, demonstrating that the defense of the homeland does not rest solely on systems or soldiers, but on citizens who view vigilance as a virtue. Because in a democracy, the sky may belong to all, but its safety begins with each of us.

Chapter 12 – The Citizen's Shield: Building a Culture of Awareness

Section 1: From Fear to Familiarity: Normalizing Responsible Vigilance

"Eternal vigilance is the price of liberty.", Thomas Jefferson. "Awareness is not fear, it's faith in action." Homeland Security Field Motto, 2024.

It looked like a videography drone. White shell, GoPro mount, gentle hover. It drifted near the outdoor venue where two officers exchanged vows in full dress uniform. Guests smiled. Kids waved. Then the signal dropped. The drone nosedived, but not by accident. Its payload wasn't a camera. The casing held a modified flashbang. No one was hurt. But the message was unmistakable: you don't need a battlefield to be targeted anymore.

The first step in national defense doesn't happen in a war room. It occurs when an ordinary citizen looks up, sees something unusual, and decides to say something. For too long, the public has viewed homeland security as someone else's job, a government function, a military specialty, a professional class of defenders. But in an age where small drones can carry considerable consequences, every household has a role. Defense today begins not with the question "Are we ready?" but with "Are we aware?" Awareness is the connective tissue between freedom and security. It transforms fear into focus and citizens into sentinels.

Since 9/11, Americans have lived in a state of cautious vigilance. However, over two decades later, that vigilance has given way to fatigue. Threats have evolved, quieter, cheaper, and airborne, while public perception has remained stagnant. The problem is psychological: people tune out danger when it feels distant or when they feel powerless. That's why awareness campaigns often fail; they warn but rarely empower. We don't need a culture of anxiety. We need a culture of agency. Instead of "watch for threats," the message should be: "You are part of the defense network." That's not fearmongering. That's nation-building.

Drones symbolize both freedom and fragility. They've democratized flight, empowered creators, and revolutionized emergency response. But they've also blurred the line between hobby and hostility. The paradox is simple: the same device that delivers medicine can also deliver mayhem. A drone hovering over a wildfire may be a firefighter's tool or a hazard grounding aircraft. That ambiguity demands awareness, not suspicion of every pilot, but understanding of what's normal and what's not. When citizens learn to distinguish between benign and suspicious activity, they turn uncertainty into intelligence.

Part of building awareness is dismantling fear through familiarity. People fear what they don't understand, and drones remain misunderstood. Public education must begin with transparency: what drones are capable of, the regulations governing their flight, and the rights and responsibilities of both operators and observers. Imagine if every local library hosted quarterly "Know Your Airspace" sessions, where law enforcement, hobbyists, and residents learned together. That's how vigilance becomes culture, through conversation, not confrontation. Knowledge doesn't just disarm ignorance; it disarms fear.

In a democracy, citizens are the distributed sensors of the Republic. Millions of eyes on the ground are more powerful than any single radar in the sky. The "See Something, Say Something" campaign remains relevant, but it must evolve for the drone era. The next step is "See Smart, Share Safely." This means observing responsibly without harassment or speculation, reporting accurately using proper channels and apps, and responding calmly without panic or misinformation. When awareness becomes collaborative, it scales. Each citizen becomes a sensor node in a human network of national resilience.

Awareness is passive; literacy is active. A literate public doesn't just notice drones; it understands context. Drone literacy encompasses understanding FAA airspace classifications, visual line-of-sight limitations, the distinctions between commercial, recreational, and unauthorized (rogue) use, and how to report airspace violations without confrontation. Just as we teach financial literacy to prevent exploitation, drone literacy prevents manipulation, both by negligent pilots and by those with malicious intent. When knowledge is distributed, power is diffused.

Awareness campaigns come and go with budgets. Cultures endure through ownership. Communities must take awareness personally, integrating it into schools, town halls, and local emergency programs. Think of drone awareness like CPR training: something everyone should know, but hope never to use. Cultural change happens when responsibility becomes routine. It's when a citizen no longer says, "That's not my job," but, "That's my neighborhood."

Resilience isn't built on technology alone; it's rooted in mindset. Where fear isolates, shared purpose unites. Community institutions, whether civic, cultural, or ethical, can play a critical role in reframing awareness. It's not paranoia to prepare; it's stewardship in action. When local leaders speak to the values of vigilance, they often succeed where policy memos fail, because they tap into moral duty rather than just procedural compliance. In this way, awareness shifts from suspicion to solidarity. Preparedness isn't a sign of distrust. It's a quiet agreement between people and principle: that safety is everyone's responsibility, and resilience begins long before the emergency. When shared values drive awareness, not fear, threats lose their grip.

Fear is reactive. Awareness is reflective. One paralyzes; the other prepares. To build a vigilant culture, communities must normalize mindful alertness, the kind that observes without becoming overly preoccupied. Think of it like defensive driving: you don't panic at every car, but you stay alert enough to act if one swerves. Homeland security works the same way, constant readiness without constant alarm. Vigilance becomes sustainable only when it's grounded in calm, not chaos.

In 2023, after a series of drone incursions near local events, Nashville Metro Police launched Project SkyWatch, a public-private partnership between citizens, businesses, and emergency management. Residents were trained to identify legitimate commercial activity and report anomalies through a community app. The result: a 40 percent reduction in false drone reports and a significant interdiction of an illegal surveillance drone during a concert. Success didn't come from technology alone; it came from trust. People stopped fearing the sky and began to understand it.

Fear thrives in silence; confidence grows through repetition. The more people engage with responsible drone use, as hobbyists, volunteers, or observers, the more normalized vigilance becomes. Imagine every high school STEM fair including a "Drone Safety & Ethics" exhibit. Imagine

PTA meetings hosting guest talks by local FAA reps. Imagine youth groups competing in community-mapping missions using drones for good, surveying flood damage, tracking forest health, and aiding first responders. When education replaces alarmism, awareness becomes second nature.

The greatest threat to public safety isn't ignorance, it's apathy. People freeze not because they don't care, but because they don't know how to respond. Awareness training must teach calm action: observe first, verify details, report through proper channels, and avoid confrontation whenever possible. Confidence replaces panic when citizens know the process. Preparedness is not paranoia; it's patriotism in motion.

At a community drone-awareness workshop, a resident asked, "What if I see a drone spying on my backyard?" The instructor smiled: "If it's hovering, call the police. If it's delivering pizza, tip well." The room laughed, and the fear began to fade. Humor disarms anxiety. Familiarity follows.

Normalizing vigilance means reprogramming perception. The drone overhead isn't always a spy; it might be a student, a surveyor, or a rescuer. When communities learn to differentiate, the sky no longer feels like a threat and becomes a shared space again. Familiarity doesn't mean complacency; it means confidence. It's the maturity to look up without panicking and the wisdom to speak up when it matters. Because the greatest danger isn't the drone itself, it's a citizenry that stops paying attention.

In the next section, *"Educating the Next Generation: Drone Literacy in Schools,"* we'll explore how to institutionalize awareness through education, transforming classrooms into incubators of civic responsibility and ensuring that the next generation learns not just how to fly drones, but how to live wisely under them. Because the surest defense of tomorrow begins with the lessons we teach today.

Section 2: Educating the Next Generation: Drone Literacy in Schools

"Education is the most powerful weapon which you can use to change the world.", Nelson Mandela. However, "If we don't teach them what to look for, we can't expect them to protect what we've built.", CSM, Homeland Defense Command Briefing, 2025

Every nation teaches its children how to read, write, and reason, but few teach them how to observe. In an age when threats can hover silently at 400 feet, that omission is no longer sustainable. Drone literacy must become as fundamental as digital literacy. Students already live in a world of constant connectivity; it's time they learn that awareness isn't surveillance, it's stewardship. When young people understand the skies, they inherit both opportunity and accountability. They stop seeing

300

airspace as abstract and start recognizing it as shared terrain, one that demands both curiosity and conscience.

For decades, STEM, comprising Science, Technology, Engineering, and Math, has driven modern education. But the acronym needs an upgrade. It's time to expand it into STRONG: Science, Technology, Resilience, Observation, Navigation, and Governance. This approach integrates not only technical skill but also civic strength. A student who can code a drone should also be familiar with airspace law. A teen who films with a quadcopter should understand privacy rights. And every graduate should leave school knowing that safety and innovation must rise together. Drone literacy isn't just about flying; it's about foresight.

Effective education programs rest on three interlocking pillars: knowledge, ethics, and action. Students must understand the various types of drones, their operations, and the associated legal parameters governing their use. They must respect privacy, data collection, and the responsible use of data. And they must know how to report unsafe or suspicious behavior. When all three are taught together, students don't just learn the mechanics; they understand the meaning behind them. That's how a culture of vigilance grows: through curriculum, not crisis.

Drone literacy belongs not only in science labs but in civics classrooms. A unit on "Airspace Awareness" can connect history, geography, and technology, teaching students how America's freedom of flight carries the responsibility of restraint. Imagine middle schoolers learning how

301

drones assist firefighters and farmers, what FAA Part 107 means for local communities, and why data from the sky can both help and harm. They'd leave not just smarter, but more secure. Every lesson becomes an act of defense.

Teachers are the unsung sentinels of democracy. They shape perception long before policy does. By integrating drone education into their lessons, they can demystify technology and defuse fear. A geography teacher might assign students to map their town's infrastructure vulnerabilities using drone footage simulations. A civics teacher might lead a debate on the topic of privacy versus safety. A theology teacher might explore moral responsibility in human invention. Each lesson builds both intellect and integrity. Education, when done right, doesn't just teach knowledge; it trains conscience.

No school can do this alone. Technology companies, drone manufacturers, and defense innovators must become stakeholders in civic education. Partnership programs can provide classroom simulators and drone kits, as well as guest lectures from FAA officials or engineers, and sponsored competitions focused on "Drone Safety and Innovation." This type of collaboration serves both the mission and the market. When industry helps educate, it earns trust, and trust is the currency of tomorrow's airspace. The private sector doesn't just build drones; it builds the reputation of responsible innovation.

Character education once focused on the virtues of honesty, respect, and diligence. Now, it must include digital and aerial ethics. When students learn that flight carries moral weight, they begin to see freedom not as a license but as a legacy to be cherished. Teaching respect for airspace mirrors respect for others; both require discipline over desire. A generation that learns to fly responsibly will also learn to lead responsibly.

Faith-based schools have a unique opportunity to integrate drone literacy through a lens of stewardship and responsibility. When Scripture commands us to "guard the flock" and "keep the watch," it's not just metaphorical; it's operational. Students can learn that awareness is a form of service, that vigilance is an expression of love for community, and that technology must always serve creation, not the other way around. A prayer before launch and reflection after flight remind students that even innovation has an altar.

To replace fear with familiarity, schools should allow students to take risks. Supervised drone programs teach technical skills while reinforcing responsibility and accountability. Search-and-rescue simulations enable students to locate mock survivors using thermal imaging. Environmental studies can track erosion, reforestation, or water levels. Disaster mapping projects can partner with local emergency management offices to enhance their capabilities and improve response times. When young people use drones for good, they view technology as a tool, not a threat. And when they see themselves as capable stewards, vigilance becomes second nature.

Imagine "Sky Labs," school-based hubs where students, teachers, and citizens share data, discuss drone sightings, and practice responsible reporting. These labs could host Drone Awareness Days featuring live demonstrations, public workshops for parents on drone safety and privacy laws, and student task forces that assist emergency agencies during natural disasters. Such labs transform schools into engines of local resilience, connecting classrooms to command posts and fostering a transition from learning to leadership.

Awareness is an inheritance. Every generation passes its knowledge, values, and vigilance to the next. Our grandparents practiced blackout drills. Our parents practiced fire drills. Our children must now practice airspace awareness. These drills shouldn't breed fear; they should foster pride: the pride of being prepared, informed, and united. Resilience isn't inherited; it's taught.

In 2024, the Texas Department of Education piloted a Drone Literacy Framework in 25 public schools. Students learned the basics of drone operation, FAA rules, and airspace safety. The curriculum also included "Ethics in Innovation," prompting debates on AI, privacy, and surveillance. Within a year, participating schools reported a 50 percent increase in students pursuing STEM fields and a measurable rise in civic engagement. Educators called it "teaching curiosity with conscience."

During a classroom Q&A, a student asked, "Can we use drones to spy on our teachers?" The instructor replied, "You can, but they'll use them to grade you faster." The room erupted in laughter. And in that moment, a lesson was learned: technology always cuts both ways, depending on who holds the controls.

When drone literacy becomes part of the education, awareness shifts from being reactionary to reflexive. Students begin to see themselves as contributors to security, not consumers of it. The goal isn't to raise pilots, it's to raise protectors. The next generation must inherit not just airspace, but accountability. Because the sky doesn't belong to the government alone, it belongs to everyone who cares enough to look up.

In the next section, *"Local Leadership: Sheriffs, Mayors, and Citizen Liaisons,"* we'll explore how municipal leaders serve as the connective tissue between federal strategy and neighborhood security, the point where policy meets people and vigilance becomes local, because national defense begins where the streetlight ends.

Section 3: Local Leadership: Sheriffs, Mayors, and Citizen Liaisons

"All politics is local, and so is all security." And so, "The first responder to any crisis is the community itself."

The strength of a nation is measured not by its capital, but by its counties. National security may be drafted in Washington, but it's executed in Waco, Wichita, and Woodland Park. Federal policies provide frameworks; local leaders provide follow-through. In the evolving drone landscape, sheriffs, mayors, and emergency managers are no longer just administrators; they're airspace guardians. They translate national directives into neighborhood realities, ensuring that awareness doesn't stop at the county line. Every drone sighting, every report, every local ordinance adds up to the defense of the Republic.

County sheriffs remain among America's most trusted officials, elected by the people, accountable to the people, and often the first to respond when something unusual crosses the sky. Unlike federal agencies with vast jurisdictions, sheriffs operate where trust is personal and response is immediate. They are perfectly positioned to lead Community Drone Response Protocols (CDRPs), quick-reaction frameworks that train deputies and civilians on how to identify, report, and coordinate with federal airspace authorities. The sheriff's badge must now symbolize more than law enforcement; it must represent layered defense, the intersection of constitutional authority and technological adaptation. When sheriffs lead awareness, communities follow confidence.

If sheriffs uphold the law, mayors inspire it. Mayors are the moral tone-setters of their towns; their words define what vigilance looks like without fear. A mayor who speaks intelligently about drone safety or resilience reframes the conversation from one of suspicion to one of stewardship. By holding quarterly Community Resilience Forums, mayors can unite businesses, schools, churches, and local media around shared preparedness goals. Good leadership doesn't amplify anxiety; it anchors awareness. A well-informed mayor can transform a city of spectators into a coalition of participants.

Policy often fails not because of poor design, but because of poor delivery. Citizen liaisons, trained volunteers who connect communities with local agencies, bridge that gap. They serve as the neighborhood's eyes, ears, and voice in emergency management. Their job isn't surveillance; it's communication. Programs like CERT (Community Emergency Response Teams) already exist. Still, they must evolve into CERA (Community Emergency and Reconnaissance Awareness), where volunteers are educated about drone behavior, reporting channels, and information sharing. When citizens know the process, panic turns into partnership.

Successful local defense is built on trust networks, not just tech networks. Consider Colorado Springs' forming a SkySafe Coalition, where the police department partnered with NORAD and the FAA to develop a rapid drone-response protocol, where they trained city staff, schools, and businesses on reporting procedures and reducing false alarms by 60 percent. In Maricopa County, the Eyes Above Program established a hotline and app for citizens to report suspicious drone activity safely, while also providing public education on FAA rules. Participation doubled within a year. Tampa's Community Shield Initiative brought police, fire, and faith leaders together for monthly resilience roundtables, reinforcing the idea that security begins with shared accountability. The formula is simple: communication plus collaboration equals credibility.

Churches, synagogues, and mosques have historically served as sanctuaries, places of safety during both natural disasters and moral crises. Today, they can also be hubs of awareness. When a pastor reminds his congregation to stay alert, it reaches hearts that policy memos never will. Faith leaders can integrate vigilance into ministry, turning "love thy neighbor" into "protect thy neighborhood." A partnership between law enforcement and faith-based organizations multiplies trust. People listen to those who pray with them before they'll trust those who patrol near them. That's why local leadership must engage both pulpits and precincts.

Proximity, the closeness of local leaders to their communities, gives them a unique advantage in terms of credibility and trust. Citizens are more likely to report suspicious drone activity to someone they know personally. That's why sheriffs and mayors must remain visible, relatable, and reachable. Town halls, social media Q&As, and local radio segments all maintain a conversational tone rather than a confrontational one. Security loses momentum the moment it loses connection.

Local governments should establish baseline certification for drone-awareness responders, similar to CPR or first aid. Recommended modules include Airspace Awareness 101, which recognizes the distinction between legal and unauthorized flight; Incident Reporting Procedures, which understand who to contact and what data to collect; Public Communication Etiquette, which prevents panic through clear messaging; and Faith and Community Integration, which engages local networks to foster support and build trust. Trained citizens and staff multiply response capacity while reducing the risk of misinformation or overreaction. Preparedness should be the new patriotism.

Local leaders must also recognize that awareness saves money. A false alarm that grounds an airport or cancels a concert costs far more than a prevention program ever will. By investing in community education, mayors and sheriffs can avoid costly overreactions and protect both public safety and economic continuity. Awareness is not an expense; it's an efficiency. Every dollar spent on readiness is ten saved on recovery. Innovative governance treats resilience as infrastructure.

Leaders set the emotional temperature of their communities. When fear spikes, people need composure more than commentary. In a crisis, a calm leader isn't passive; he's persuasive. A trembling community needs to see steadiness embodied. A proverb often quoted in the military says, "Calm is contagious." The same principle applies to civic life. A mayor who speaks reassurance steadies nerves; a sheriff who models readiness steadies hearts. Faith-driven leadership transforms anxiety into action.

Local media outlets are powerful amplifiers, but they can either build awareness or breed hysteria. Mayors should cultivate media relationships before a crisis occurs. Regular briefings, public service messages, and shared vocabulary, using "unauthorized drone activity" instead of "drone threat", ensure that the narrative educates, not alarms. Information discipline is a form of defense. The enemy thrives in confusion; leadership thrives in clarity.

Local agencies often hold valuable information, such as airspace violations, emergency response patterns, or suspicious trends, that remains inaccessible to the public. Creating Community Intelligence Networks (CINs) enables the safe and transparent sharing of data among law enforcement, city governments, and residents. By releasing summarized reports, such as "five unauthorized drone flights detected this month", agencies reinforce awareness while maintaining operational security. Transparency builds trust; trust builds vigilance.

At a city council meeting about drone ordinances, a resident stood up and said, "If a drone flies over my house again, I'm shooting it down." The sheriff smiled and replied, "Then we'll both be filing reports." The laughter eased the tension and opened the door for dialogue. Humor, when used wisely, is a civic pressure valve. It disarms defensiveness while reinforcing respect for the law.

Public office is not just administrative, it's covenantal. A mayor's oath, a sheriff's badge, both are pledges of stewardship. When leaders treat their roles as sacred trusts, vigilance becomes contagious. Communities mirror the integrity of those who lead them. Faith-based ethics, transparent governance, and consistent communication are not optional extras; they are the scaffolding of local resilience because no federal plan can replace moral leadership on the ground.

The next evolution of leadership is mentorship. Every mayor, sheriff, and emergency manager should identify and train successors, young citizens who will inherit the responsibility of local vigilance. Programs like Junior Resilience Corps or City Cadet Ambassadors can inspire civic service from an early age. The future of security depends not only on systems, but also on their successors. Leadership that reproduces itself ensures continuity of conscience.

In the next section, "Technology for Good: Drones in Rescue, Disaster Relief, and Research," we'll explore how the same tools that threaten can also protect, how drones, when guided by principle and partnership, become instruments of mercy, innovation, and recovery. Because technology itself isn't moral; it's the hands that command it that define its purpose.

Section 4: Technology for Good: Drones in Rescue, Disaster Relief, and Research

"Every tool can build or break, depending on the hands that hold it." For all the danger drones can pose, they also represent one of the greatest humanitarian innovations of the century. In the right hands, a drone is not a weapon; it's a winged witness of compassion. It can see

what the human eye cannot, go where rescuers cannot, and deliver what time will not allow. When guided by purpose instead of profit, these small machines become extensions of mercy. The same technology that can disrupt can also deliver. The same sky that shelters threats can also shelter hope.

Search and rescue has always been a race against the clock. In disasters, whether hurricanes, wildfires, or earthquakes, time becomes the most valuable resource, and drones have proven to be the most versatile ally. Equipped with infrared sensors, they locate heat signatures of missing persons through smoke and debris. Fitted with loudspeakers, they direct stranded survivors toward safety. When roads are washed out, they carry medicine, food, or communication relays. In 2023 alone, drones contributed to over 1,200 documented life-saving missions across the United States. Every flight was a testimony to human ingenuity serving human dignity.

When fires engulfed parts of Maui, local responders faced collapsed infrastructure and blinding smoke. Traditional helicopters were grounded by poor visibility, but drones filled the gap. Within hours, emergency crews deployed small UAVs equipped with LiDAR and thermal imaging capabilities to map the damage, identify survivors, and assess the fire's progression. The real miracle wasn't the technology; it was the coordination. Firefighters, volunteers, and drone hobbyists worked together under a unified command, proving that when citizens become partners, resilience becomes reflex. The mission's unofficial motto captured the moment perfectly: "The sky fought back, for us."

Beyond search and rescue, drones have revolutionized the logistics of disaster relief. Where trucks can't go, drones glide. Where communication towers fail, drones step in. Organizations like the American Red Cross, Team Rubicon, and GlobalMedic now integrate drones into every major operation. In 2024, after devastating floods in Kentucky, drone teams delivered over 10,000 pounds of supplies to cut-off communities, a feat that would have been impossible by road for nearly two weeks. What once required convoys now takes coordinated flight paths. Technology, when disciplined by compassion, becomes a force for grace.

Nature, too, has found its defenders in the drone age. Researchers now use UAVs to monitor endangered species, map coral reef bleaching, and study glacial retreat with precision once thought impossible. In Alaska, scientists track caribou migrations without disturbing the herds. In California, drones map wildfires in real time, allowing crews to predict spread patterns and save lives. In Florida, marine biologists use underwater drones to monitor coastal health after hurricanes. Each mission reaffirms a profound truth: stewardship is not static; it's adaptive. The same skies that once carried destruction now carry devotion to creation itself.

Faith communities have also embraced drones as tools of mercy. Churches and NGOs in Africa utilize drones, also known as UAVs, to deliver vaccines to remote villages. Mission organizations deploy drones to map flood plains before disaster strikes, preparing relief efforts before the first drop of rain. In Malawi, a partnership between local parishes and UNICEF created "mission corridors", pre-approved drone routes for medical aid, protected under prayer and policy alike. It's faith meeting flight, and the Gospel taking wings. The drone, once feared as a symbol of surveillance, becomes an emblem of service.

What makes drones unique and challenging is their dual-use nature. The same payload mount that can carry a weapon can also transport water bottles. The same camera that can spy can also save. That duality forces society to make a moral choice every time technology evolves. The solution is not to ban innovation, but to bend it toward benevolence. As one Air Force chaplain said during a humanitarian airlift, "God doesn't fear our inventions; He judges our intentions."

Universities across the nation have incorporated drones into their scientific research. From Purdue's agricultural monitoring systems to MIT's autonomous search algorithms, research institutions are turning drones into platforms of discovery. But perhaps most importantly, they're using them to teach responsibility. Courses such as "Ethics of Autonomous Systems" and "Drone Governance and Policy" prepare the next generation of engineers to strike a balance between innovation and integrity. That's how academia becomes the new arsenal of ethics. Knowledge itself becomes a form of defense.

Police and fire departments are increasingly relying on drones for situational awareness, but transparency is crucial to maintaining trust. When appropriately managed, drones reduce risk to officers and civilians alike. When misused, they erode privacy and legitimacy. Departments like Chula Vista Police in California now operate under the "Transparency and Trust" model, where every drone deployment is logged, publicly reported, and reviewed by a civilian board. The result? Reduced crime, faster response times, and increased community support. Accountability is the new airworthiness certificate.

The Old Testament is full of messengers, divine couriers sent to deliver hope or warning. In some ways, drones have become the modern metaphor for that mission: to see, to reach, to serve. When guided by compassion, they embody a distinctly spiritual ethic, eyes in the sky for hands on the ground. Technology doesn't replace divine purpose; it amplifies it. As James 2:17 reminds us, "Faith by itself, if it is not accompanied by action, is dead." Each rescue mission, each research flight, each humanitarian delivery is an act of faith with propellers.

Empowering communities to use drones for good also means teaching restraint. Every new capability must come with a renewed conscience. Public-private partnerships should include not only training in technical operation, but in moral application. The question must always be asked:

who benefits, and who bears the risk? When innovation asks for permission before acting, it transforms from disruption to devotion. The mark of a mature society isn't how advanced its tools are, but how wisely they are used.

At a drone safety workshop, a firefighter joked, "If my drone crashes, it's a malfunction. If your kid's drone crashes, it's evidence." The room laughed, but the message landed; context matters. That's the point: drones are only as dangerous, or as decent, as the intent that guides them.

It's poetic, really, the same airspace that once symbolized vulnerability now represents possibility. Every responsible drone operator becomes a guardian of that transformation. We can't eliminate risk, but we can elevate purpose. We can't predict every misuse, but we can prepare every moral response. Technology doesn't define us; stewardship does. And every act of responsible innovation is a quiet victory for civilization itself.

In the next section, *"The Shield We Share: Practical Blueprint for Civic Resilience,"* we'll conclude this chapter by outlining actionable frameworks for how communities, agencies, and individuals can institutionalize vigilance through education, training, partnerships, and faith. Because awareness without action is observation, but awareness with organization is defense.

Section 5: The Shield We Share: Practical Blueprint for Civic Resilience

"The strength of a nation is not in its walls but in its will." George Washington (paraphrased)
"Security is everyone's job, complacency is everyone's risk.", DHS Readiness Directive, 2025.

Awareness without structure drifts into noise. Action without coordination collapses into chaos. Resilience lives in the space between, where understanding meets organization. This blueprint isn't theoretical; it's tactical. It's a guide for communities ready to move from concern to capability, from spectatorship to stewardship. It's how ordinary Americans become the invisible armor of national defense. Because the shield that protects this nation isn't forged in Washington, it's built in living rooms, classrooms, and county halls.

Every community must build a network before the crisis. A Civic Resilience Council (CRC) should be established locally to bring together city leaders, law enforcement officials, educators, faith-based representatives, and volunteer groups under a single umbrella. Their mission is to coordinate awareness, response, and recovery related to incidents involving drones or airspace. Quarterly workshops on airspace awareness, direct communication channels with state and federal agencies, and simplified public briefings become the backbone of readiness. When communities prepare together, they respond more quickly and feel less fear.

Every person with a smartphone becomes part of the national observation grid. Apps like B4UFLY, AirHub, and local reporting portals should be promoted and standardized for practical use. Citizens don't need to become drone experts, just good witnesses. Record, don't engage. Report, don't speculate. Respond calmly, not chaotically. When millions of citizens share structured vigilance, the collective situational awareness outpaces any adversary's innovation. It's crowd-sourced defense, powered by discipline, not paranoia.

Drone literacy must evolve into Civic Technology Education (CTE), a permanent addition to public and private school curricula. Schools should host annual Sky Safety Weeks that engage students through simulations, drone demonstrations, and ethics workshops. Colleges and universities must integrate Public Technology Ethics into general education. A resilient nation begins with a literate citizenry. Knowledge neutralizes fear faster than policy ever can.

Resilience isn't a government monopoly; it's a marketplace of integrity. Private-sector partners can contribute equipment, expertise, and innovation to community programs in exchange for trust, the most valuable public currency. Local governments should formalize agreements with drone manufacturers, telecommunications companies, and logistics firms for mutual support during emergencies. Imagine a city where Amazon's delivery drones, the county's emergency drones, and the Red Cross's rescue drones all operate under a unified protocol. That's not science fiction, it's civic fusion.

Faith communities are the emotional infrastructure of resilience. When disaster strikes, people often turn to their pastors before seeking government policies. Churches, mosques, and synagogues should appoint Faith Liaison Officers, trained volunteers who bridge moral encouragement and material assistance. They organize watch programs, coordinate aid, and offer calm communication during crises. Faith gives vigilance a heartbeat. It turns readiness into reassurance.

Every community should conduct annual Blue Sky Exercises, local simulations that test readiness for drone-related emergencies, just as hurricane or fire drills test natural-disaster plans. Scenarios might include unauthorized drone activity over a stadium, loss of communications infrastructure due to aerial interference, or coordinated delivery of emergency supplies via drone fleet. Each drill must end with an After-Action Reflection, not just "what went wrong," but "what did we learn about our unity?" Preparedness is not fear; it's fellowship under pressure.

Information spreads faster than any drone can fly. That's why communities must establish pre-vetted communication chains. Primary alerts should come via emergency apps or local radio. Secondary updates must be shared through verified social media channels managed

319

by CRC public affairs officers. Tertiary communication should flow through faith and neighborhood networks for those offline. A rumor can undo what readiness has built. Discipline in speech is as essential as precision in strike.

Every successful shield must be inspected. CRC councils should issue quarterly transparency reports outlining drone-related incidents, response effectiveness metrics, and lessons learned. Publishing this data builds trust, and trust is the bedrock of sustained vigilance. Secrecy invites skepticism; transparency invites participation.

Federal grants for community preparedness often expire or get reallocated. Thus, to make resilience sustainable, local governments must create Resilience Trust Funds, small, recurring budget allocations for awareness campaigns, training, and technology maintenance. Money spent on prevention multiplies itself in recovery savings. Preparedness is fiscal responsibility disguised as patriotism.

Resilience culture must feel rewarding, not restrictive. Communities should celebrate readiness achievements the way they celebrate parades and holidays. Awards for Citizen Watch Leader of the Year or Drone for Good Student Project turn vigilance into civic pride. The goal is not to burden citizens with responsibility, but to honor them for it. Fear divides; pride unites. Resilience should feel like a sense of belonging, not a burden.

When every household takes ownership of awareness, every city becomes a fortress without walls. The shield we share is invisible but invincible, made of facts, trust, and fellowship. It is strengthened by each citizen who reports with courage, each leader who listens with humility, and each community that stands together when the sky turns uncertain. Resilience is not an emergency plan; it's a civic covenant, an agreement between the governed and the governing that vigilance will never again be outsourced.

The American experiment has always depended on participation. From the minutemen to the modern responder, security has never been the privilege of the few; it's been the duty of the many. In a republic, vigilance is the rent we pay for freedom. But it's a rent paid in attention, not anxiety, through education, empathy, and engagement. A nation that watches together will never fall apart.

During a community debrief, a volunteer said, "We're not paranoid, we're just professionally alert." The sheriff replied, "Good. That's cheaper than rebuilding a bridge." The laughter eased the room, but the truth remained: awareness saves money, lives, and legacies.

The final step of this blueprint isn't on paper; it's in practice. Resilience is not a checklist; it's a character trait. It grows stronger every time someone chooses involvement over indifference. The citizen's shield isn't made of steel or software. It's built from trust, truth, and tenacity, the three elements no adversary can counterfeit. Because when a nation stands together in awareness, it becomes unbreakable in spirit.

In the next chapter, "*Above and Beyond: The Future We Choose,*" we rise from blueprints to vision, synthesizing everything we've learned about technology, morality, and civic duty into one defining truth: that the sky will reflect not our machines, but our maturity. Because algorithms don't write the future of flight, accountability does.

Chapter 13 – Above and Beyond: The Future We Choose

Section 1 – The Fork in the Flight Path: Innovation or Implosion

"The future depends on what we do in the present." Mahatma Gandhi. And so, *"The sky isn't our ceiling; it's our mirror."* CSM Sheldon A. Watson.

It was a routine high school football game during the playing of the national anthem. The crowd is on its

feet. Then a tiny drone buzzed the flagpole, too close, too fast. Security hesitated. Was it filming? Was it hostile? The coach's brother, a National Guardsman, spotted the trajectory. With muscle memory from a training exercise, he tackled a trash can, launching its lid at the drone like a discus. It hit. The drone shattered. The play made the local paper. But the question lingered: why did a civilian need to improvise a counter-drone defense at a school?

Every generation faces a defining choice, a moment when progress demands principle and speed tests discipline. For us, that moment is now, hovering between innovation and instability. The drone revolution began as a marvel and matured into a necessity. However, without coordinated governance, the same systems that connect us can also compromise our security. The question is no longer how high we can fly; it's how steady we can stay once airborne. Innovation without intention isn't progress, it's propulsion without direction.

Every great invention carries two flight plans: one of potential, one of peril. Fire cooked and consumed. Electricity enlightened and executed. The atom healed and harmed. Drones and AI have become the latest reflection of that duality: efficient, scalable, unstoppable, yet capable of disruption when unmoored from restraint. Technology reveals character. When power outruns responsibility, societies drift into turbulence of their own making.

We now live in an age where machines learn faster than institutions legislate. AI pilots react in milliseconds, while regulations often take years to crawl through the review process. This gap between capability and accountability widens with each innovation. It's not malicious, it's mechanical. However, it still breeds risk. The longer we delay aligning ethics with engineering, the more our progress outpaces our preparedness. History won't remember how fast we advanced, but how responsibly we managed acceleration.

The sky offers both promise and permission. It tempts nations and corporations to act first and regulate later, to chase capability before consequence. But ambition without boundaries invites chaos. Unchecked innovation turns the airspace into a free-for-all, where hobbyists, corporations, and adversaries compete for dominance with no shared set of rules governing their activities. We've seen this movie before, in cyberspace. And like any domain without discipline, it breeds disorder before stability. The sky deserves better governance than guesswork.

Tomorrow's leadership will not be defined by who can invent the most, but by who can govern the best. The next great American advantage will be ethical agility, the ability to adapt technology within frameworks that balance both commerce and conscience. That requires what I call the Three A's of Airspace Leadership: Awareness, a complete understanding of technological reach and risk; Accountability, clear ownership of consequences and compliance; and Adaptability, the will to evolve policies as fast as the tools they regulate. These aren't bureaucratic virtues; they're survival instincts for modern democracies.

For two centuries, America's strength has rested on its ability to pair invention with restraint. We built the internet, but also the laws to govern it. We weaponized flight but also created the FAA to protect it. Now, as drones and autonomous systems expand, that balance is being tested again. If innovation continues to sprint ahead while oversight jogs behind, we'll find ourselves leading a race with no finish line and no rules. Power alone doesn't define progress; proportion does.

Neglect rarely makes headlines, but it always writes history. If we fail to adapt governance to the pace of technology, the erosion won't be sudden; it'll be subtle. Privacy is lost by increments, accountability blurred by automation. Public trust has diminished due to constant exposure and unequal enforcement. The actual threat to national security is not attack, it's apathy, a slow surrender to convenience over caution: every unreported incident, every unreviewed algorithm, every unexamined data policy compounds into vulnerability. Complacency is how complex systems collapse, quietly, predictably, and avoidably.

Aircraft maintain altitude through lift and balance; nations maintain theirs through integrity and foresight. To stay aloft, we need three stabilizers: Transparency, open communication between government, industry, and the public; Trust, earned through consistent, ethical performance; and Truth, data-driven policy, not ideology-driven reaction. Lose one stabilizer, and governance begins to wobble. Lose all three, and the airspace becomes ungovernable. Moral altitude isn't rhetoric; it's a national flight requirement.

At a defense tech symposium, an engineer joked, "Our drones make decisions faster than Congress forms committees." The laughter was instant and uneasy. The truth stung because it exposed the gap between innovation and institution. We can't afford that lag anymore. The future won't wait for parliamentary procedure.

To every technologist, policymaker, and entrepreneur: your fingerprints are on the trajectory of history. Build with caution. Test with conscience. Deploy with discipline. If capability becomes our only compass, the destination will always be regret. The mark of leadership isn't how fast we climb, but how safely we land.

We now stand at a fork in the flight path of civilization. One direction leads to structured innovation, guided by ethics, equity, and oversight. The other veers into reckless autonomy, efficiency without empathy, control without conscience. Machines won't make the decision; humans will, choosing accountability or abdicating it. The sky isn't empty, it's expectant. And what fills it next will reveal what kind of society we truly are.

In the next section, "From Policy to Principle: Embedding Ethics in Law," we'll examine how to transform good intentions into enforceable frameworks, ensuring that innovation remains a servant of civilization, not its substitute. Because the laws we write today will determine whether the next century of flight becomes a triumph of governance or a monument to neglect.

Section 2 – From Policy to Principle: Embedding Ethics in Law

"Good laws make it easier to do what's right and harder to do what's wrong."

And so, "Innovation without governance is simply risk on autopilot."

Every new technology begins as a disruption, matures into dependence, and ultimately demands discipline. But policy rarely evolves fast enough to keep pace with that curve. Drone and AI systems are perfect examples; their growth has outpaced the regulatory ecosystem built to contain them. We are operating 21st-century tools under 20th-century frameworks. FAA Part 107 was never designed for swarm coordination, AI-assisted navigation, or cross-border drone operations. The gap between law and life has become a liability. We now face a national challenge that isn't about technology itself; it's about governance keeping pace with the rapid advancements in technology.

The reasons for policy inertia are structural, not malicious. Congress legislates slowly by design. Federal agencies move methodically by necessity. Industry innovates rapidly due to profit incentives. When these three forces collide, deliberation, caution, and ambition, progress comes to a standstill. The result is what I call the Policy Vortex: innovation outpaces oversight; incidents spark reactive regulation; regulation lags, creating public frustration; industry resists, fearing profit loss; and Congress retreats, fearing political backlash. And so the cycle repeats, every new drone, every new software patch, every new loophole. We can't govern the airspace of the future with policies written for the skies of the past.

Laws can't predict innovation, but they can guide its intent. What's missing in drone governance isn't more pages of regulation; it's a principle-based framework that anchors innovation to ethics. This requires three shifts. First, from restriction to responsibility: move from "what cannot be done" to "what must be done safely." Regulation should clarify accountability, not suffocate advancement. Second, shift from a reactive to a proactive approach: develop adaptive legislation that evolves in response to emerging threats, utilizing rolling updates and public-private advisory boards. Third, from compliance to conscience: make transparency and safety prerequisites for operation, not afterthoughts added after an incident has occurred. A responsible drone ecosystem doesn't fear regulation; it depends on it.

Just as the National Security Strategy defines our defense posture, the United States needs a National Drone Doctrine. This guiding framework integrates the commercial, civil, and defense uses of unmanned systems under shared principles. This doctrine should establish ethical standards for development and deployment, define federal, state, and local responsibilities in shared airspace, mandate transparency in data collection, retention, and transmission, and outline escalation and enforcement mechanisms for misuse. Without such a doctrine, we are flying blind, not for lack of technology, but for lack of direction. A nation that leads the airspace must also legislate it.

Policy cannot rely on goodwill alone; it must encode moral intent into measurable law. The following principles can serve as a foundation for modern airspace legislation. First, an Accountability Clause: every drone must be traceable to an operator through Remote ID systems, no exceptions for private or commercial entities. Second, a Data Integrity Requirement: any collection of imagery or telemetry data must meet defined privacy and retention standards, with criminal penalties for misuse. Third, Autonomy Thresholds: define limits for machine autonomy; human oversight must remain the fail-safe in all operations with kinetic or surveillance potential. Fourth, Dual-Use Safeguards: manufacturers of dual-use systems must certify anti-tampering safeguards to ensure civilian drones cannot be easily weaponized. Fifth, Liability in Code: software developers share responsibility for breaches that enable unlawful drone behavior when resulting from negligent design. These five tenets turn ethics from aspiration into architecture.

Federal control can't cover every rooftop. That's why the National Drone Doctrine must empower states and municipalities to enact local ordinances aligned with national principles, much like traffic laws vary by region but follow consistent standards. Examples include altitude zoning to define safe vertical corridors over schools, hospitals, and power stations; noise and privacy standards to protect residential zones from persistent noise and surveillance; and community reporting protocols to streamline citizen participation in incident reporting. Federalism works best when it enables consistency without crushing initiative. Washington can set the tone, but the counties will set the tempo.

Instead of viewing the industry as the regulated, treat it as the regulator's partner. Technology can help enforce the very standards it risks violating. For instance, AI-based geofencing can prevent unauthorized drone entry into restricted areas. Blockchain systems can record drone activity logs, ensuring transparent flight histories. Cloud-based monitoring can automatically report anomalies in real-time to local authorities. The same ingenuity that creates risk can create resilience, if appropriately guided. Regulation should not only respond to innovation; it should leverage it.

Other nations have already recognized that airspace ethics must evolve faster than airspace economics. The European Union's Aviation Safety Agency implemented a risk-based regulatory model tailored to each operation type, not to aircraft size. Israel combines oversight of national defense with civilian integration, ensuring continuous data flow between agencies. Japan created the UAS Traffic Management network, enabling the safe coexistence of drones, manned aircraft, and automated air routes. Each of these models emphasizes adaptability and data sharing. America's framework should not imitate; it should integrate. We must learn from the world, not trail it.

A law unenforced is an invitation, not a deterrent. The Department of Homeland Security, the FAA, and the Department of Justice must

331

coordinate enforcement authority under a single operational protocol. This includes unified training for local law enforcement on drone response, standardized penalties and escalation paths, as well as the sharing of threat intelligence between federal and state agencies. Currently, overlapping jurisdictions dilute effectiveness. A cohesive enforcement framework ensures both accountability and consistency. The airspace is shared; enforcement must be, as well.

No regulation survives without legitimacy. Public buy-in depends on two elements: clarity and transparency. Citizens should be able to understand who regulates what, how to report misuse, and how their privacy is being protected. A complex policy is an unread policy, and unread policies breed mistrust. Trust is built when people can see themselves in the system that governs them. If government, industry, and the public communicate using the same vocabulary, the airspace will no longer feel alien; it will feel like home.

At a Senate briefing, a staffer asked, "How many agencies regulate drones?" The witness replied, "All of them, depending on who you ask." The laughter that followed wasn't amusement; it was exhaustion. Policy clarity isn't about creating new agencies; it's about unifying existing missions. America doesn't need another acronym; it needs alignment.

The ultimate goal is not just to control drones but to define the character of the airspace itself, to make it predictable, transparent, and principled. That means codifying ethics as infrastructure. Predictability: legal changes should be straightforward and foreseeable, allowing businesses in the industry to innovate with confidence. Transparency: public access to regulatory data builds legitimacy. Proportionality: punishments must fit the degree of risk created, not just the nature of the violation. When law, logic, and leadership operate together, safety stops being reactive and becomes reflexive.

In the next section, *"The New American Airspace: A Shared Stewardship Model,"* we'll explore how the United States can operationalize this ethical framework, uniting citizens, corporations, and government under a cooperative model that balances innovation with public safety. Because the sky won't protect itself, it's only as strong as the people, policies, and principles that guard it.

Section 3 – The New American Airspace: A Shared Stewardship Model

"Airspace belongs to everyone, and therefore, it's everyone's responsibility." Thus, "Shared domains demand shared discipline."

For much of aviation history, airspace was treated as a vertical highway, owned, managed, and policed by federal authority. But the drone revolution shattered that model. The sky has become a shared environment: part commerce, part security, part public trust. Ownership has evolved into stewardship. In this new model, the FAA no longer owns the airspace; it orchestrates it. States no longer merely enforce; they integrate. And citizens are no longer spectators; they are stakeholders. Airspace management is no longer a government monopoly; it's a national partnership.

A shared stewardship model rests on three pillars: collaboration, coordination, and communication. Collaboration involves government agencies, private firms, and civic organizations co-developing standards that maintain safety scalability. Instead of regulation versus innovation, it becomes regulation through innovation. Coordination ensures the sharing of real-time information across jurisdictions, from FAA command centers to local sheriff's offices. Seamless interoperability prevents duplication and enables rapid response. Communication anchors public trust through transparency, from open data portals to citizen-reporting apps. Together, these pillars form the framework of modern airspace democracy.

Public-private partnerships have long played a crucial role in constructing roads, bridges, and energy grids. Now, they must create something less visible but more vital: trust in the sky. The next generation of PPPs should focus on technology integration, shared platforms for monitoring, tracking, and incident response. They should offer joint training and certification programs to standardize drone education across public and private operators. They must establish unified incident response protocols that incorporate commercial expertise into emergency operations. These partnerships aren't just cooperative ventures; they're the backbone of national resilience. When public trust meets private innovation, governance gains lift.

A theoretical framework is of little value without operational clarity. To translate shared stewardship into real-world systems, America needs a Unified Airspace Operations Center, a digital clearinghouse for all civilian, commercial, and government drone data that enables predictive analytics and shared situational awareness. Regional Airspace Coordination Hubs, managed at the state level, must connect local agencies to federal networks for synchronized enforcement. Community Air Observers, civilian volunteers trained in airspace awareness, should be deployed like neighborhood watch programs, with a focus on reporting anomalies or hazards. Each component builds redundancy into the system, so that when technology falters, people step in to fill the gap.

The industry doesn't just create the technology; it defines the pace of its adoption. That's why corporate actors must accept a role not as mere participants, but as co-governors of public trust. Companies must establish internal ethics boards to evaluate potential misuse scenarios and address them effectively. They should publish clear, accessible data-use policies that explain how collected information is stored and protected. They must also offer public accountability portals that track drone operations for commercial fleets. Innovation that hides invites suspicion. Innovation that reports earns legitimacy. The companies that lead with integrity will define not just market dominance, but moral authority in the airspace economy.

Mayors, sheriffs, and emergency managers will play the frontline role in this ecosystem. Federal agencies can issue guidance, but it's the locals who will handle first contact, the drone crash, the privacy complaint, and the suspicious sighting. Local governments need both the authority and the resources to act decisively. Federal grants and technical support must be matched with local readiness, including trained personnel, effective communication networks, and robust public engagement. When Washington and Wichita speak the same airspace language, everyone wins.

The new airspace does not rely on asphalt, but on information. Real-time data is the runway that allows drones, aircraft, and regulators to coexist safely. To make that work, we need a National Airspace Data Infrastructure, a network that standardizes data formats, secures transmission, and anonymizes citizen information. That infrastructure must be secure against cyber manipulation, accessible to federal, state, and local agencies, and designed with ethics in mind, incorporating built-in privacy compliance. Data is no longer just a record; it's a responsibility.

Universities are the laboratories of legitimacy. They can model ethical frameworks, test new airspace management tools, and train the next generation of drone professionals. Academic institutions should be designated as Centers of Airspace Governance, think tanks where policy, law, and engineering intersect. Through simulations, legal clinics, and industry partnerships, they ensure that innovation remains evidence-based and ethically anchored. In short, academia must move from analysis to application.

For shared stewardship to endure, the public must see a return on its participation. That means visible safety improvements, clear communication, and tangible community benefits. Public reporting should not feel like snitching; it should feel like service. A community that sees its vigilance reflected in policy will protect that policy with pride. Transparency breeds participation. Participation builds legitimacy. Legitimacy sustains resilience. That's the civic dividend of shared stewardship.

No airspace exists in isolation. The United States must lead internationally by example, developing exportable governance standards while reinforcing domestic autonomy. A shared stewardship model works only if our allies adopt similar standards of transparency and accountability. International coordination reduces cross-border threats and standardizes enforcement, particularly against transnational actors who use drones for illicit purposes. Global interoperability begins with national consistency.

At a summit on shared airspace, one panelist said, "We're not trying to control every drone, just the ones that don't listen." The moderator replied, "So... most of them?" The laughter cut the tension, but the truth remained: command in the sky isn't about domination, it's about cooperation.

The measure of success for shared stewardship won't be how few incidents occur, but how swiftly and smartly we respond when they do. Resilience is readiness multiplied by coordination. It's what happens when silos give way to systems, when data, policy, and people move in rhythm instead of reaction. The sky doesn't need more ownership; it needs orchestration. And that orchestration begins with a shared understanding: no single entity can defend what the entire nation occupies.

In the next section, *"The Long View: Preparing for What Autonomy Brings Next,"* we'll look beyond today's policies and into the horizon, exploring how AI-driven autonomy, swarm systems, and machine decision-making will reshape both our opportunities and our obligations. Because the airspace of the future won't wait for us, and the only way to lead it is to be ready before it arrives.

Section 4 – The Long View: Preparing for What Autonomy Brings Next

"The future doesn't arrive suddenly; it accumulates in the blind spots of the present." And so, "We taught machines to think faster than us, but not yet to think better."

We've made machines that can outthink our reflexes, but not our conscience. One of those corners is above us now, expanding with every drone that takes flight. Autonomous systems are no longer theoretical; they're operational. What began as pilot assistance has evolved into pilot replacement. Drones now patrol borders, deliver supplies, and conduct precision mapping with minimal oversight. The next generation will make independent tactical and operational decisions based on real-time data streams. The future airspace won't be managed by people watching screens; it will be handled by code that learns, predicts, and reacts faster than its designers.

The benefits of autonomy are undeniable. Machines don't fatigue, panic, or disobey. They process information in milliseconds, respond uniformly, and operate in environments too hostile or complex for humans to manage. Autonomous systems can conduct disaster reconnaissance without endangering rescuers, manage synchronized delivery networks across entire cities, and patrol borders with persistent coverage and zero downtime. In terms of efficiency and reach, autonomy outperforms human capacity. The promise is precision, decisions made cleanly, without ego, fear, or fatigue. But the very qualities that make autonomy powerful also make it perilous.

A perfect system doesn't fail often, but when it does, it fails. Autonomy compresses error into critical points of dependency: code, connectivity, and control. When an algorithmic network acts incorrectly, there's no pause button. The response time is measured in microseconds, long enough to watch it happen, too short to stop it. This is the paradox of progress: we gain speed at the expense of sovereignty. The more we delegate decision-making, the less we understand what drives it. The challenge isn't building more intelligent systems; it's keeping them interpretable.

Traditional command models rely on the OODA Loop, which consists of the following steps: Observe, Orient, Decide, and Act. Autonomous systems now execute that cycle millions of times per second, without human involvement. That shift creates three urgent questions. First, can we trust the data-feeding machine's judgment? Second, who sets the limits of acceptable action? Third, who owns the outcome when autonomy errs? Without answers, we risk trading human error for mechanical indifference, a dangerous exchange in national defense or civil governance.

Machine learning relies on data, but data reflects human behavior, and human behavior is inherently complex. Algorithms trained on imperfect datasets will replicate those imperfections at scale. Autonomy magnifies both virtue and bias. A flawed model doesn't discriminate; it just executes. If we don't embed accountability into the design, we'll be debugging ethics after deployment. That's a repair job no one can afford.

The rise of drone swarms marks the next phase of autonomous evolution, hundreds of machines acting as a single, cohesive entity. Swarm intelligence offers advantages: adaptability, redundancy, and resilience against single-point failure. But it also challenges human comprehension. Commanders can't control a swarm; they can only set intent and parameters. After that, the system adapts on its own, learning as it goes. That's not command, it's coexistence. And coexistence with code demands a new leadership mindset, one focused on governance, not guidance.

Every defense innovation in history has begun with tactical optimism and ultimately led to strategic consequences. Autonomy will be no different. In 2025, over 40 countries reported active development of autonomous or semi-autonomous combat drones. Few have established legal or ethical frameworks governing their use. That gap creates what strategists call the Autonomy Gray Zone, a realm where action is fast, attribution is fuzzy, and accountability is elusive. The next flashpoint in global security won't be who launches first; it will be who takes responsibility after the launch.

The merging of autonomy and cyber creates a new form of risk, systems that not only operate independently but can also be collectively hijacked. A compromised algorithm doesn't just fail; it becomes a weapon against its creator. In a fully autonomous network, a single vulnerability can spread across thousands of platforms in seconds. Defending autonomy will require as much cybersecurity discipline as it does airspace management. Every line of code is a potential breach. The next frontier of warfare won't be fought for territory; it will be fought for control of logic.

Traditional accountability relies on visibility, which enables tracing decisions to their respective decision-makers. Autonomous systems blur that line. When a machine acts autonomously, who answers? The developer who wrote the code? The company that deployed it? The commander who approved it? Or the system itself, which technically made the decision? Without established accountability frameworks, autonomy and responsibility diffuse faster than data processing occurs. Governance must shift focus from the originator of orders to the creators of autonomous capabilities.

The phrase "human in the loop" has become the moral safety valve of autonomy, a promise that people will always have the final say. But in reality, speed is outpacing supervision. To remain relevant, human oversight must become proactive rather than reactive. We must design for human-on-the-loop systems, where oversight is embedded, not appended. That means predictive ethics, algorithms designed to self-diagnose when decisions approach moral or operational boundaries. Autonomy needs not just limits, it needs literacy.

Governance will depend on a workforce that understands both the code and the consequences. That requires new cross-disciplinary training that combines engineering, law, public policy, and behavioral science. The next generation of policymakers must be fluent in systems logic, not just statutes. Defense universities, civilian agencies, and private think tanks must work together to develop a National Autonomy Readiness Curriculum that prepares leaders to manage technology that doesn't wait for permission. Competence is the new compliance.

Autonomy will also rewrite the economic order. Automation will create efficiencies but displace traditional labor, shifting national economies toward data management, AI integration, and systems oversight. Jobs will evolve from operators to auditors, verifying systems rather than executing them. Those nations that invest early in workforce retraining will dominate both innovation and ethics. The lesson is clear: autonomy rewards the prepared and punishes the passive.

Autonomy doesn't exist in a vacuum; it exists in competition. The question isn't whether nations will adopt it, but how they'll do so responsibly. The challenge ahead is to establish shared norms before crisis demands them. Just as nuclear arms treaties codified restraint in the 20th century, autonomy will require 21st-century equivalents, agreements rooted not in disarmament but in data discipline. If we fail to define acceptable behavior in autonomous conflict, we'll repeat the mistakes of every arms race before, only faster.

During a closed-door forum, a general quipped, "We used to worry about pilots falling asleep. Now we worry about machines waking up." The room laughed, then nodded. Because it wasn't paranoia, it was perspective. Autonomy doesn't replace humanity. It magnifies it, for better or worse.

Preparing for autonomy isn't about resisting it; it's about managing it with foresight and restraint. That balance defines mature nations. The United States can either react to autonomy or shape it by building guardrails before the first crash, not after. Because autonomy is not just about flight, it's about direction. If we lose the ability to steer, we lose the right to lead.

In the final section, *"The Closing Reflection: The Sky Isn't Empty; It's Waiting for Direction,"* we'll bring together every theme, innovation, governance, ethics, and civic responsibility, into one cohesive message: that the future of flight is not a race between machines and humans, but a test of whether progress can still serve purpose. Because the sky isn't empty, it's listening. And what we send into it reflects who we are.

Section 5 – The Closing Reflection: The Sky Isn't Empty; It's Waiting for Direction

"We shape our tools, and thereafter our tools shape us.", Marshall McLuhan.
"The sky isn't empty, it's a reflection of the discipline, duty, and direction we put into it," CSM Sheldon A. Watson.

If this book has revealed anything, it's that the sky is no longer just atmosphere; it's an archive. Every flight, every signal, every algorithm becomes part of a collective record of who we are and what we value. When we look up, we're not just seeing progress; we're seeing reflection. The airspace above isn't empty; it's filled with the consequences of our choices. And that's why governance, ethics, and vigilance aren't bureaucratic exercises; they're acts of national character.

We've entered an era where altitude is no longer the measure of greatness, discipline is. The drone revolution, the rise of autonomy, and the proliferation of airborne systems have prompted us to rethink what freedom itself means. Can liberty survive when every move can be tracked? Can democracy remain strong when decision-making shifts from people to programs? Can innovation endure if trust erodes? These aren't technical questions; they're civic ones. The answers depend less on code and more on conscience because the speed of our systems will never compensate for the steadiness of our judgment.

Every generation inherits a frontier. Ours happens to fly. We can't slow technology, nor should we. But we can shape the rules that govern it and the principles that guide it. That's the core of civic responsibility in the modern era, not to resist change, but to refine it. Whether in the Pentagon, a start-up lab, or a local town hall, the question remains the same: will we use our tools to serve people, or will we train people to serve the tools? The future won't be written by innovation alone, but by intention.

From the drafting of the Constitution to the launch of Apollo, America's power has never been just technical; it's been ethical. Our success has come not from doing everything we could, but from deciding what we should. Restraint has always been our secret weapon, the discipline to act with authority rather than impulse. That's what made the Republic durable in every domain: land, sea, space, and now airspace. The sky doesn't test our technology; it tests our maturity.

We often think of governance as a hierarchy, but it's actually a continuum, one that connects engineers to policymakers, citizens to soldiers, coders to commanders. Each plays a role in securing the same sky, just from different altitudes. The engineer ensures the system functions. The policymaker ensures it's fair. The citizen ensures it's trusted. The leader ensures it's used with wisdom. When those roles operate in alignment, we get more than safety; we get stability. That's the definition of stewardship in the modern age: shared accountability for shared airspace.

Technology doesn't stop at national boundaries, and neither does responsibility. Every drone flight, every algorithmic update, every innovation reverberates globally. The way we govern our skies will influence how others govern theirs. If we build a model of integrity, others will replicate it. If we allow chaos, it will multiply. America's leadership in this new domain won't be decided by who has the best drones, but by who has the best discipline. The future belongs to those who act with foresight, not just force.

During a discussion on drone safety, a senior engineer said, "Our biggest risk isn't machines taking over, it's humans taking a day off." The audience laughed, but the point was clear: systems don't fail because of code; they fail because of complacency. The same rule applies to nations. The moment we stop paying attention, we stop leading.

In the old world, power meant possession of territory, weapons, or wealth. In the new world, power means precision of data, direction, and discernment. Our airspace isn't a battlefield; it's a balance sheet of national discipline. Every drone in the sky, every law passed on the ground, every decision deferred or made defines our position in history. The question for the next century isn't who controls the sky, but who deserves to.

Command and control will always have their place in governance, but conscience and collaboration will define its future. The next era of leadership won't be about issuing orders; it'll be about earning alignment. Leaders who understand the balance between authority and empathy, innovation and restraint, will hold the real advantage. Because the future isn't automated, it's accountable. And accountability, once lost, is more complicated to restore than any downed aircraft.

The sky has always been a symbol of aspiration, limitless, untamed, full of possibility. But now it's also a covenant, a contract between those who innovate and those who inherit. Every flight must honor that covenant, not just in compliance, but in character. The sky may belong to everyone, but the responsibility to preserve it belongs to each of us. That's not policy, it's principle.

We must also recognize that the sky is no longer a passive backdrop; it's an active domain. It listens, it records, it reacts. It's a sensor field, a delivery corridor, a surveillance grid, and a strategic theater. And in that complexity lies a simple truth: the sky reflects the systems we build and the values we embed within them. If we fill it with noise, it will echo confusion. If we fill it with discipline, it will echo trust.

The closing reflection, then, is not a conclusion; it's a call to continuity. The sky isn't empty because it's filled with what we send into it: our machines, our data, our ambition, and our intentions. It's not empty because it mirrors our progress and exposes our priorities. It's not empty because it waits for us to choose between control and chaos, between leadership and drift. The sky isn't the final frontier; it's the first reflection. And what it shows back to us will depend entirely on the direction we give it.

This is not just a technological imperative; it's a leadership imperative. The systems we've built are impressive, but they are not self-governing. They require judgment, calibration, and care. They need leaders who understand that progress without purpose is just acceleration. They require citizens who understand that vigilance is not paranoia, it's participation. And they require institutions that understand that trust is not a given, it's a currency earned through transparency and restraint.

The evolution of airspace governance will involve contributions not only from engineers but also from educators, ethicists, emergency management professionals, and members of the public. It will be shaped by those who ask not just what a drone can do, but what it should do. It will be shaped by those who understand that the sky is not a blank canvas, it's a shared responsibility.

We are entering an age where altitude is easy, but direction is hard. Where lift is automated, but leadership is manual. Where the tools we've built can either elevate our values or expose our vulnerabilities. The choice is ours. And it begins not with a launch, but with a mindset.

In the next chapter, "The Leadership Imperative: Guiding Through Uncertainty," we shift our focus from systems to souls, from how the sky is managed to who manages it. Because technology can automate decisions, but it can't inspire discipline. And in a world where command has become code, authentic leadership, leadership that remains the last human advantage, remains paramount.

Chapter 14 – The Leadership Imperative: Guiding Through Uncertainty

Section 1 – The Vacuum of Voice: When Leaders Defer to Data, Followers Drift Toward Doubt

"Data can guide decisions, but only leaders can define direction.", CSM Sheldon A. Watson.

Game day. Forty thousand fans. A drone buzzed overhead, disrupting the streaming signals across the stadium. Advertisers panicked. Fans were unable to scan QR codes or order drinks. The drone wasn't armed, but it was curious. It was collecting metadata. The tech company contracted for stadium security hadn't accounted for signal spillovers. Their solution: install more towers. But the problem wasn't coverage, it was control. When private entities manage public spaces, who gets to define what is considered "safe"? And when no one speaks up, who gets to decide?

In every era of disruption, one phenomenon repeats: when uncertainty rises, leaders retreat into analysis. They wait for better data, more clarity, one more report, until the moment for decision has already passed. Technology has made this reflex worse. With metrics for everything and

351

accountability for nothing, the modern leader can hide behind dashboards rather than provide direction. But when leaders defer to data instead of discernment, organizations lose confidence. Numbers may show precision, but silence signals doubt. And in any chain of command, military or civilian, doubt spreads faster than any signal.

Modern decision-makers drown in information. Every sensor, every satellite, every spreadsheet demands attention. But leadership isn't about collecting more data; it's about filtering the right kind. A thousand metrics can't replace one clear message. Too often, leaders confuse data comprehension with command competence. They believe understanding the analytics equals understanding the situation. It doesn't. Data describes, it doesn't decide. Leadership is the act of translating information into intention, something algorithms can't replicate.

The greatest casualty of automation isn't jobs, it's voice. When systems become self-sustaining, leaders risk becoming spectators. This is what I call the vacuum of voice: a condition in which authority still exists but presence does not. Meetings continue, memos circulate, metrics update, but no one is actually leading. The vacuum doesn't appear suddenly; it forms quietly. It starts with a leader delegating a judgment call to "the model." Then, a department defers to "the data." Soon, people stop asking why altogether. When the human voice leaves the room, the machine hum becomes the soundtrack of governance.

352

There's a dangerous myth circulating in both government and industry: that data is neutral. It's not. Data reflects the biases of its collectors, the assumptions of its coders, and the priorities of its users. When leaders treat data as absolute truth rather than contextual input, they surrender judgment to the past, because all data is, in essence, a reflection of the past. By the time it's analyzed, it's already describing what was, not what is. Leadership, on the other hand, is future-oriented. It's not about knowing everything, it's about knowing enough to decide, and then having the courage to act. A leader's authority doesn't come from certainty; it comes from conviction.

Automation has made it easy to outsource not just tasks, but responsibility. When something fails, it's convenient to blame the system. "We followed the model." "The algorithm made the call." "The data didn't predict this." These phrases sound safe but hollow. They signal the erosion of ownership, and ownership is the cornerstone of leadership. Machines can optimize, but they can't justify. They can measure success, but they can't define it. When decisions become mechanical, accountability becomes optional.

Leadership is a burden precisely because it requires imperfect information. The best leaders make calls with incomplete data and stand by them anyway, not recklessly, but responsibly. Command presence isn't about charisma or rank. It's about the willingness to speak when everyone else is waiting for certainty. In combat, silence from the top kills momentum. In civil governance, it kills morale. A leader's hesitation sends one message louder than any order: we're not sure. And once that uncertainty reaches the front line, whether it's a platoon or a policy office, discipline starts to erode.

Modern leaders fear being wrong more than they fear doing nothing. Social media, instant scrutiny, and perpetual metrics have made decisiveness a liability. But leadership isn't about being right all the time; it's about being responsible every time. A wrong decision, corrected, still moves the mission forward. No decision at all leaves everyone waiting, and waiting is how organizations fail. The paradox of leadership in the digital era is that we've never had more data, yet we have never been more afraid to use it.

Followers don't need leaders to be flawless; they need them to be present. They need a steady voice that says, "I've got this," even when the system doesn't. In a crisis, trust doesn't flow from analytics; it flows from accountability. When people see their leaders step forward rather than hide behind processes, confidence becomes contagious. The human voice, calm, clear, and consistent, remains the most stabilizing force in any environment. And in an era where people hear from screens more than from supervisors, that voice has never mattered more.

Technology has expanded every leader's reach, but diluted their presence. Emails, dashboards, and AI decision aids have replaced conversations. But leadership cannot be automated. It's a contact sport. It requires eye contact, not just analytics; empathy, not just efficiency.

The digital chain of command still needs human links, people willing to say the hard things, make tough calls, and take responsibility for the outcomes. That's the difference between authority and influence. Authority can be programmed; influence must be earned.

In 2024, a multi-state cyberattack crippled parts of the U.S. logistics grid. Data streams flooded in from dozens of agencies, each waiting for definitive analysis before taking action. By the time a decision was made, 18 hours had passed, and with it, millions of dollars in supply losses. The failure wasn't technical; it was psychological. Everyone had information, but no one had initiative. That event became a lesson taught at the National Defense University: "Data didn't fail us. Delay did."

During a crisis debrief, an officer once said, "Sir, the data's still coming in." The commander replied, "So's the enemy, make a call." The room laughed, but the message landed: clarity beats completeness every time. Leaders who wait for perfect data rarely lead at all.

The solution to the vacuum isn't more dashboards, it's more dialogue. Leaders must reassert their presence, in tone, in timing, and in truth. That means speaking early, even when uncertain, and acknowledging risk without surrendering resolve, translating complexity into clarity. When leaders reclaim their voice, followers regain their confidence. And when confidence returns, data once again becomes what it was meant to be, a tool, not a crutch.

Leadership begins where algorithms end, at the intersection of judgment, timing, and moral courage. The machines can show the way, but they cannot lead the way. In an age obsessed with analysis, the true differentiator is still voice, the leader's steady hand on the throttle, saying, "This is where we go next." Because when leadership goes silent, doubt takes command. And doubt, once in charge, never leads anyone home.

In the next section, *"Character in Command: Integrity, Humility, and the Courage to Decide Amid Ambiguity,"* we'll explore how leadership strength isn't about knowing everything, but about anchoring judgment in values when clarity is scarce. Because when data reaches its limit, character becomes the compass.

Section 2 – Character in Command: Integrity, Humility, and the Courage to Decide Amid Ambiguity

"Character is what you do when no one's watching; leadership is what you do when everyone is waiting." Thus, "Rank gives authority; integrity gives permission."
CSM Sheldon A. Watson.

And in moments of ambiguity, when the fog doesn't lift and the data doesn't clarify, character becomes the only reliable compass.

Leaders like to believe that decisions become easier with experience. The truth is, experience just makes you more aware of the weight. The burden doesn't shrink, it sharpens. And when visibility drops to zero, character fills the space that clarity leaves behind. It's the ballast that keeps a leader steady when the mission loses shape. Whether in combat zones or corporate offices, ambiguity isn't an obstacle; it's the environment. And those who can't lead through uncertainty shouldn't lead at all.

Integrity isn't about compliance; it's about consistency. It's doing what's right when no one can verify it, and standing by it when everyone can. In the Army, integrity is the invisible chain of command. It links the leader's decision to the follower's trust. If that chain breaks, the mission collapses, no matter how well-planned it is. Integrity ensures that decisions are made for purpose, not for convenience, optics, or politics. And that's a distinction lost in many modern institutions: too many leaders seek validation instead of verification. When integrity is intact, followers don't need to understand every decision; they just need to trust the intent behind it.

Humility is often mistaken for weakness. It's not. In leadership, humility is the confidence to acknowledge one's limits, the discipline to listen, and the wisdom to make adjustments. The best leaders I've known never acted like they had all the answers; they just made sure answers got found. Humility doesn't shrink command; it strengthens it. It reminds teams that leadership isn't a one-way transmission but a shared responsibility. In a room full of egos, humility is the rarest rank. It allows a leader to hear warning signs before they become headlines. Humility doesn't mean hesitation; it means situational awareness of self.

357

Also, courage isn't just about combat. It's about standing alone in a meeting when everyone wants to compromise the standard. It's about making the unpopular call because it's the right one. It's about telling the truth, even when it threatens the outcome. Modern leadership requires more moral courage than ever before. Every decision is broadcast, judged, and archived for future reference. The margin for error is razor-thin. Yet courage remains the deciding factor between leaders who react and those who rise. Without it, integrity stays quiet and humility turns into avoidance. Courage activates both; it gives the leader a voice when others go silent.

Integrity, humility, and courage are not separate traits; they form a system. Integrity gives direction. Humility gives awareness. Courage gives execution. A leader lacking any one of these is like a pilot missing an instrument:

Without integrity, there's drift.

Without humility, there's blind confidence.

Without courage, there's paralysis.

Together, they keep a leader centered, regardless of the chaotic conditions.

True character isn't revealed by ceremony or command; it's shown by stress. When the radios fail, when the plan unravels, and when accountability evaporates, that's when leadership reveals its true nature.

Under pressure, weak leaders default to control. Strong leaders default to connection. They speak directly, own up to mistakes, and stabilize morale through their presence alone. Command isn't about eliminating chaos; it's about absorbing it without transmitting panic. That's the essence of character under pressure: the ability to stay calm when others look to you for permission to fall apart.

Today's leaders face an impossible expectation: flawless performance in real time, under full public scrutiny. This obsession with perfection discourages authenticity and punishes honesty. But real followers don't want perfect leaders; they want predictable ones. Leaders whose values don't change when outcomes do. Perfectionism creates paralysis. Character creates confidence. A leader who can admit error and correct course earns more loyalty than one who hides behind metrics and jargon. The best commanders I've seen weren't the ones who never stumbled; they were the ones who stumbled forward.

There's a phrase often used in command briefings: "Sir, with all due respect, we don't know." A weak leader hears defiance in that sentence. A strong one hears opportunity. Humility allows leaders to surround themselves with truth-tellers rather than with echo chambers. It ensures blind spots get exposed before they get exploited. When leaders create an environment where "I don't know yet" is an acceptable phrase, they foster innovation and initiative instead of fear. The best units, teams, and organizations are those where everyone feels responsible for thinking, not just for obeying.

Decisiveness is not recklessness. It's the skill of making a call based on principle when the data is incomplete and the timeline is unforgiving. That kind of decision-making separates managers from leaders. Managers wait for policy; leaders define it. Ambiguity is where leadership earns its paycheck. Anyone can execute a plan; it takes character to make one in the dark.

In 2012, during a deployment in Kandahar, a platoon was caught in an ambush near an unsecured airstrip. The radios were jammed, the maps outdated, and visibility down to meters. The platoon sergeant, not the senior officer, decided to reposition and hold perimeter using an improvised strategy based on terrain memory. It wasn't textbook, but it worked. When asked later how he made the call, he said, "I didn't need more data. I needed to do what was right for my people in that moment." That's leadership in ambiguity: not guessing, but grounding.

Technology scales efficiency; character scales trust. An organization can survive without resources, but not without credibility. Every leader is a multiplier or a divider. Integrity multiplies confidence. Humility multiplies cooperation, courage multiplies initiative. Without them, no system or structure can hold. Character turns compliance into commitment, and that's what keeps nations, units, and institutions from fracturing under stress.

During a leadership course, a junior officer asked, "How do you stay calm when everything's going wrong?" The instructor said, "I drink a cup of coffee so strong it's legally considered a morale booster." The room laughed, but he added, "Calm isn't caffeine, it's character." Every leader knows that moment when calm isn't natural, but necessary.

Leadership isn't taught through slides or slogans; it's transferred through example. People don't follow plans; they follow presence. They mimic what they see tolerated and emulate what they see rewarded. That's why the leader's conduct isn't personal; it's cultural. Every act of integrity or indifference sets a precedent that ripples across ranks and years. In the long run, your replacement will follow your lead. So teach them wisely, because legacy is just leadership that outlived its author.

Character isn't a talking point; it's a survival tool. It won't make decisions easier, but it will make them defensible. When everything else is uncertain, the mission, the metrics, the future, integrity, humility, and courage remain nonnegotiable instruments. They're the only navigation tools that never need recalibration.

In the next section, "*Mentorship in the Machine Age: How Leaders Teach Judgment When Algorithms Tempt Shortcuts,*" we'll explore how modern leaders can pass down intuition, patience, and ethical reasoning in a world that increasingly values speed over wisdom. Because good data builds systems, but good mentors build successors.

Section 3 – Mentorship in the Machine Age: How Leaders Teach Judgment When Algorithms Tempt Shortcuts

"Mentorship isn't about creating followers; it's about training thinkers who can lead without permission.", CSM Sheldon A. Watson.

"We can automate the process, but we can't automate the principle." That's the leadership paradox of our time. As systems grow smarter, the need for wisdom grows sharper. And wisdom, unlike data, doesn't scale; it's transferred.

Before technology began predicting our next move, people learned by watching others make hard decisions. The old model of mentorship, shoulder-to-shoulder, real-time, consequence-driven, built judgment from observation and repetition. It taught not just what to do, but why. Now, those lessons are being replaced by tutorials, dashboards, and chatbots. The human transfer of wisdom, the part that teaches discernment, is disappearing. Leadership without mentorship becomes technical, not tactical. It produces operators, not thinkers; managers, not mentors. And when that happens, institutions don't evolve; they just age.

Technology's greatest temptation is convenience. Algorithms reward immediacy, not maturity. These algorithms offer certainty without context, quick answers without the slow work of understanding. That's a problem. In leadership, shortcuts don't save time; they hinder development. When younger leaders grow up trusting systems over instincts, they stop learning how to think under pressure. And when a crisis hits, when the system freezes, fails, or feeds false data, the leader's mind becomes the only processor left. Mentorship keeps the processor sharp.

A mentor's purpose isn't to provide instructions; it's to model discernment. To show how to remain calm when others panic, and be curious when others assume. To expose the process of thinking, not just the product of it. That's why mentorship in the digital era must focus on two pillars:

Judgment under ambiguity.

Ethics under pressure.

These can't be learned from manuals. They're taught through proximity, conversation, and the courage to admit mistakes in front of those learning from you. Mentorship isn't about creating replicas; it's about producing leaders capable of improving on your flaws.

Younger generations now grow up believing every problem has an answer hidden in a search bar. They've been conditioned to expect immediate clarity and low friction. But leadership requires the opposite: patience, reflection, and tolerance for uncertainty. Mentorship reintroduces friction, and friction builds depth. Good mentors don't give solutions; they give questions that force reflection. Because leadership isn't memorization, it's adaptation. A system can recall data. Only a human can apply wisdom.

Technology can simulate knowledge, but it can't replicate presence. A mentor's tone, timing, and demeanor during tough decisions communicate lessons no lecture ever could. The act of standing next to someone while they make their first consequential call, that's mentorship in its purest form. It's where confidence is transferred not through words, but through composure. Presence teaches poise. Poise sustains judgment. And judgment is the one skill that can't be downloaded.

In an age of constant noise, mentorship serves as a moral shield. It teaches discernment against both external pressures and internal shortcuts. When mentees are guided to slow down, analyze ethically, and act deliberately, they become resilient against the chaos of automation. Mentorship builds a filter, one that catches ego before it becomes arrogance, and ambition before it becomes recklessness. That filter doesn't come from an app; it comes from someone who's already made the mistakes you're about to make.

Every great mentor teaches at least three lessons, directly or indirectly:

- Be predictable in principle, not routine in behavior.
- Ask questions that matter more than the answers.

- Remember that silence is often the most honest form of feedback.

These lessons never expire, even when technology changes every 18 months. They remind leaders that while tools evolve, human nature doesn't.

One of the most overlooked crises in modern institutions isn't turnover; it's knowledge drain. As seasoned leaders retire or transition, their replacements often inherit the seat but not the insight. Mentorship bridges that gap. Without it, organizations end up repeating lessons that were already paid for, sometimes in blood, sometimes in budget. In the military, we refer to it as losing continuity. In the private sector, it is often referred to as a 'losing culture.' In both, it means the same thing: forgetting what was already learned the hard way.

Mentorship isn't a one-way street. The best leaders listen as much as they teach, not because they're uncertain, but because they're smart. Younger professionals bring innovation, speed, and a fresh perspective. Experienced leaders bring judgment, restraint, and historical awareness. Together, they balance the system, ambition meeting experience, energy meeting discipline. This kind of reverse mentorship builds resilience on both sides. The senior leader stays current. The junior leader learns caution without cynicism. It's not hierarchy, it's harmony.

Technology has shortened every feedback loop, except the one that matters most: maturity. Character takes time. Mentorship protects that time from being wasted on premature promotion or impulsive reactions. A mentor doesn't slow progress; they ensure it's sustainable. They remind future leaders that speed without depth leads to burnout, and ambition without boundaries leads to downfall. Mentorship doesn't rush excellence; it refines it.

As part of a significant federal task force in 2023, a promising analyst developed an AI model to predict emergency logistics failures. The model was impressive until it started prioritizing efficiency over equity, diverting resources from low-income communities to high-value zones. No one caught it in time because no one mentored the analyst on the ethical layer of operational design. The lesson wasn't about coding; it was about context. Technology accelerated the mission. Mentorship could have preserved its integrity.

During a leadership seminar, a young officer once asked, "What's the fastest way to become wise?" The instructor said, "Make mistakes, and make sure I'm there when you do." That's mentorship: controlled failure with a safety net. You can't simulate that experience; you can only live it under guidance.

The best mentors don't just shape the next generation; they build continuity between them. They understand that their influence may not be felt immediately, but will echo later, in decisions they'll never witness. That's the quiet power of mentorship. It doesn't make headlines, but it prevents catastrophes. Leaders who mentor invest in resilience that outlasts their command. Because in the end, the question isn't how many people you led, it's how many you prepared to lead without you.

Mentorship is the antidote to automation. It re-humanizes leadership by teaching patience in a world addicted to speed and values in a culture obsessed with visibility. Machines may run operations, but only mentors sustain civilizations.

In the next section, *"Leading Beyond Fear: Restoring Trust Through Clarity, Empathy, and Consistency,"* we'll explore how true leadership isn't maintained through authority or analytics, but through emotional steadiness. This kind rebuilds trust when institutions falter and people begin to lose faith. Because fear may be contagious, but so is composure.

Section 4 – Leading Beyond Fear: Restoring Trust Through Clarity, Empathy, and Consistency

"A leader's calm doesn't come from knowing the outcome; it comes from controlling the tone." CSM Sheldon A. Watson.

Fear is a survival response. In battle, it sharpens instincts. In organizations, it dulls them. The difference lies in how leaders handle uncertainty. When fear spreads unchecked, whether through a platoon, a company, or a nation, logic goes offline. People stop reasoning and start reacting. In those moments, leadership becomes less about commanding others and more about stabilizing them. Good leaders don't eliminate fear; they absorb it. They act as emotional shock absorbers, absorbing the impact so that others can continue moving forward.

Uncertainty breeds speculation, and speculation breeds fear. That's why clarity, not charisma, is the leader's first weapon against chaos. Clarity doesn't mean knowing everything. It means explaining what's known, what's not, and what's being done about it. It's direct, factual, and calm, three traits that consistently outperform theatrics. During a crisis briefing, people don't need comfort; they need coherence. The most powerful sentence a leader can say is: "Here's what we know, here's what we don't, and here's what we're doing." That's the sound of confidence without arrogance.

Empathy isn't softness; it's awareness. It's understanding the room's temperature before making decisions that affect it. Leaders who ignore emotion invite rebellion disguised as compliance. When people feel unseen, they stop believing their effort matters. Empathy doesn't mean agreement; it means acknowledgment. It's pausing to understand before instructing. In one deployment, I saw an exhausted squad leader stop mid-patrol to check on a private who had mentally checked out. That five-minute conversation did more to restore operational performance than any briefing could have. Empathy isn't a detour from discipline; it's what makes discipline sustainable.

Trust doesn't come from grand gestures; it comes from predictable behavior over time. Consistency turns leadership from performance into reliability. When followers know what to expect from their leader, tone, standards, and reactions, they stop managing anxiety and start managing their tasks. Inconsistent leaders, regardless of their talent, erode confidence. Teams can adapt to stress, but not to volatility at the top. Leadership consistency is like flight stability: you don't notice it until it's gone.

In today's hyper-connected environment, fear moves faster than facts. Social media, news cycles, and public scrutiny amplify every mistake before leaders have time to correct it. That pressure tempts leaders to prioritize optics over truth, to "message" instead of communicating. But when leaders spin, followers sense it immediately. Once that trust is broken, no amount of strategy can repair it. Transparency doesn't always win applause, but it consistently earns endurance.

Every leadership situation boils down to three controllable variables:

- **Tone:** how you deliver information.
- **Timing:** when you deliver it.
- **Truth:** what you deliver.

Tone controls emotion. Timing controls confusion. Truth controls credibility. When all three align, trust compounds. It's not complicated; it's just rarely practiced.

Fear in a unit or organization isn't just emotional; it's operational. It slows decisions, stifles initiative, and turns creative thinkers into cautious bureaucrats. In a military context, fear costs time. In governance, it costs progress. Every leader must recognize that morale and mission share the same fuel: belief. If people believe in the mission and the leader's steadiness, they'll endure anything. If they doubt either, even success feels temporary.

In 2022, during a simulated air defense exercise, a radar anomaly triggered a false high-threat alert. For six tense minutes, a command team faced the unthinkable, a possible inbound strike. Two senior officers reacted differently. One froze, demanding complete data verification before speaking. The other stood up, said calmly, and laid out precise steps: "We'll verify, but until then, maintain posture, stay professional, and keep comms disciplined." When the anomaly was cleared, the team didn't remember the error; they remembered the calm. Leadership is rarely about eliminating fear; it's about directing it into discipline.

When fear dominates, noise multiplies: gossip, speculation, and finger-pointing. That's when leaders earn their pay. They must become the quietest person in the room, not by withholding emotion, but by mastering it. Command voice doesn't always mean volume; it means control. In modern organizations, calm isn't passive; it's strategic. Because in times of panic, people don't follow the plan; they follow the pulse.

370

There's a fine line between empathy and indulgence. Effective leaders care deeply, but they don't confuse compassion with compromise. When you empathize without enforcing standards, you create sympathy without structure. When you enforce without empathy, you create compliance without commitment. The balance is simple: listen to people's concerns, but lead them through them. Don't validate fear, transform it. That's leadership alchemy: turning anxiety into alignment.

During prolonged uncertainty, people stop needing good news; they need consistent updates. Predictable communication acts like heartbeat monitoring, proof that leadership is still present, still functioning. Silence breeds speculation, and speculation is the seedbed of fear. Even a 30-second check-in can reset morale for hours. Leaders often underestimate how far calm words travel, farther than fear ever does.

During a security briefing, a nervous analyst once asked, "Sir, what if everything goes wrong?" The commander said, "Then we'll have something new to learn." Laughter followed, and tension broke. The lesson wasn't flippant; it was strategic. Humor, when used with timing and tact, is leadership's pressure valve. Controlled levity reminds people that while the situation may be serious, it's still survivable.

Fear thrives in confusion, but collapses under communication. The more leaders explain, the less fear they have to feed on. That's why consistency matters: repetition turns reassurance into reality. You don't beat fear with slogans, you outlast it with steadiness. In leadership, courage is contagious, but consistency is curative.

When leaders lead beyond fear, they restore something more important than control; they regain confidence. And confidence is the currency of every mission, team, and nation. Fear isolates; clarity unites. Fear freezes; empathy mobilizes. Fear fluctuates; consistency endures. The question every leader must answer isn't "How do I eliminate fear?" but "How do I lead so people don't have to?"

In the next section, *"Legacy as Leadership: Why the Measure of Command Is Not Compliance, but Conscience,"* we'll conclude this chapter by exploring how leadership outlasts position, how it becomes a memory, an influence, and a moral blueprint for those who follow. Because a command may end, but a character leaves a lasting impression.

Section 5 – Legacy as Leadership: Why the Measure of Command Is Not Compliance, but Conscience

"Your legacy is built in the decisions no one sees and the standards no one lowers, because rank ends; influence doesn't.", CSM Sheldon A. Watson.

Every command ends, some with ceremony, others with silence. The guidon passes, the office clears, and another name takes the slot. What doesn't leave is the residue of leadership: the invisible culture you built through tone, decisions, and presence. Legacy is the shadow leadership casts long after the spotlight moves. It's measured not by plaques or promotions, but by how people behave when you're no longer there to correct them.

Many organizations mistake obedience for loyalty. They reward compliance because it's measurable. But conscience is what sustains them when orders stop coming. Compliance makes systems efficient; conscience makes them moral. The most dangerous culture isn't one of rebellion, it's one of quiet compliance, where people do what they're told but stop caring why. Great leaders don't demand conformity; they cultivate conviction. They teach followers how to think, not just what to execute. When conscience drives action, standards survive structure.

Legacy begins with an example. People copy what they see tolerated and amplify what they see honored. A single act of fairness, restraint, or courage becomes institutional memory. Long after your signature fades from documents, the story of how you led will still circulate in hallways and after-action reviews. An example is the only form of leadership that doesn't require your permission to keep teaching.

However, legacy isn't free; it comes at a cost. It means making decisions that won't be appreciated now but will be understood later. It means saying "no" to easy wins and "yes" to lasting standards. True legacy requires patience with misunderstanding. Every leader who changes culture faces resistance before reverence. If everyone agrees with you in real time, you're probably not leading; you're managing consensus.

Money and medals fade, but moral credibility compounds. It becomes institutional equity, the unseen reserve leaders draw upon when crisis hits. That's why every ethical decision matters, even the small ones. Each choice adds or subtracts from the credibility account your successors will inherit. Integrity isn't personal property; it's generational currency.

Legacy survives through mentorship. Teaching someone else to think critically, act decisively, and lead ethically ensures that your influence endures beyond your timeline. The best mentors know they're planting trees they'll never sit under. But shade doesn't appear overnight; it's built through repetition, correction, and care. Your true résumé isn't the number of missions led, it's the number of people who lead better because you were there.

Culture isn't built in policy memos, it's built in patterns. The leader who enforces standards the same way every time sends a message louder than any directive: fairness is non-negotiable. That predictability turns anxiety into trust and confusion into cohesion. When consistency becomes culture, organizations don't need constant oversight; they self-correct. That's the highest compliment a leader can earn: a team that functions ethically without supervision.

At the end of a career, the toughest reflection isn't the one in the mirror, it's the one in other people's eyes. How do they remember your tone in a crisis? Did your presence calm or complicate? Did your standards elevate or exhaust? Those answers form your legacy more accurately than any evaluation ever will. Because leaders don't write a legacy, it's written about them.

During a retirement ceremony, a junior soldier whispered, "I hope I leave half the impact he did." His buddy said, "Then you'd better start doing twice the work he did." The room laughed, but it was true. Legacy isn't luck, it's labor repeated long enough to look effortless.

Leadership that depends on a title expires with it. But influence grounded in character transcends position. When people still quote your guidance, apply your principles, or invoke your standard years after you're gone, that's command without command. That's legacy. It's the quiet authority that outlasts formal authority, the gravity that keeps a culture orbiting its values.

Every transition of command should transfer more than keys; it should transfer conscience. It's not enough to pass on processes; leaders must pass on principles. That's the moment where ethics either continue or collapse. And it depends entirely on whether you led from compliance or conviction. If your replacement inherits a framework instead of just a file, you've done your job.

Legacy is built in the decisions no one sees and the standards no one lowers. It's not about being remembered, it's about being repeated. The real measure of command is not compliance, it's conscience. Not how tightly you controlled people, but how strongly you prepared them to act right when you're not there. That's legacy: when the culture remembers your intent better than your instructions.

In the next chapter, *"The Horizon Within: Restoring Purpose in the Age of Automation,"* we close the loop of this book, returning from systems and strategies to the soul of the matter: humanity itself. Because no algorithm can define purpose, no machine can raise a family, and no automation can replace conviction. Progress may propel us forward, but only purpose keeps us from drifting.

Chapter 15 – The Horizon Within: Restoring Purpose in the Age of Automation

Section 1 – The Human Pulse: What Remains Irreplaceable When Everything Else Becomes Automated

"The machines can compute faster, but they can't care deeper." "Progress without humanity isn't progress at all, it's drift at altitude." CSM Sheldon A. Watson.

A child stood in the field, holding a kite string in one hand and a controller in the other. One toy rose with the wind. The other rose on battery life and code. His grandfather watched from a lawn chair, silent. "Why do you always make me fly the kite?" the boy asked. His grandfather smiled. "Because the drone listens to you. The kite listens to the sky." In an increasingly programmed world, some things still teach us how to feel the wind.

For all our advances, satellites that see through clouds, algorithms that anticipate behavior, machines that move without pilots, one truth holds:

the human pulse still sets the tempo of civilization. Every automated system, every digital framework, and every autonomous drone ultimately depends on the same fragile heartbeat: the people who design, deploy, and make decisions. Technology may extend reach, but it doesn't replace reason. It can amplify empathy, but it cannot manufacture it. The human pulse remains the ultimate governor of progress. And when that pulse weakens, when compassion fades, when purpose dulls, the system follows.

We often hear the phrase "the rise of the machines," as if technology were an invading species. But the real threat isn't replacement, it's relinquishment. It's when people stop engaging, stop questioning, and stop caring because systems now do it "for them." Automation tempts us with comfort: autopilot in the cockpit, algorithms in the office, artificial intelligence in the marketplace. Each promises precision and efficiency. But each dulls our moral reasoning muscles as well. The most dangerous kind of automation isn't mechanical, it's psychological. It's when humans begin to believe they're optional.

Being human in the age of automation is an act of discipline. It requires choosing awareness over convenience, reflection over reaction, empathy over efficiency. Every significant advance, from flight to fusion, has needed a human to stay present when the system said, "I've got it." Pilots still monitor the autopilot. Doctors still verify AI diagnostics. Commanders still confirm machine recommendations before launching. Because presence is not redundancy, it's a responsibility. And the moment we surrender that, we risk turning innovation into abdication.

Machines fear mistakes. Humans learn from them. That's what keeps us irreplaceable. Failure, reflection, and adaptation form the rhythm of human progress. Automation, for all its brilliance, doesn't learn; it adjusts. Because in the end, only humans transform failure into wisdom. Every breakthrough in aviation, medicine, and governance began as a human error redeemed by perseverance. That's the irony: perfection isn't what defines progress, persistence does. If the error disappears, so does evolution.

In a world obsessed with data, empathy remains the highest form of intelligence. It allows leaders to anticipate consequences that algorithms can't measure: morale, dignity, grief, and hope. Empathy is the human equivalent of predictive analysis; it sees beyond inputs and outputs into intent. And intent is where morality lives. Artificial intelligence can simulate compassion, but it can't feel it. It can draft a condolence letter, but it can't mean it. Empathy gives leadership its legitimacy. It's what turns command into connection.

We've mistaken convenience for progress. The more efficient our tools become, the less connected we seem to feel. Digital networks have made communication instantaneous, yet understanding has never been slower. Automation has freed time, yet purpose feels shorter. Efficiency without meaning creates vacancy: busy hands, empty hearts. A society that runs faster than its conscience eventually outruns its humanity. We can automate processes, but not principles.

Every system fails eventually. Circuits burn out. Networks collapse. Batteries die. But what brings the world back online is never code; it's courage, resourcefulness, and cooperation. When a power grid fails, people knock on neighbors' doors. When communications crash, first responders improvise with hand signals and instinct. When systems go dark, humanity lights the way, literally and figuratively. That's the essence of resilience: not redundancy in design, but reliability in spirit.

Automation has made life easier, but not simpler. We're connected to everyone and no one at once, speaking constantly but listening rarely. We scroll past each other's struggles and call it awareness. We outsource empathy to emojis. We confuse engagement with understanding. The human pulse isn't just biological, it's relational. When we stop making eye contact and start outsourcing compassion, society's rhythm skips a beat. Machines don't erode humanity; apathy does.

Progress tells us to accelerate. Purpose tells us to pace. The human pulse operates best with rhythm, not constant acceleration, but deliberate tempo. That's why leaders pause before speaking, parents listen before correcting, and soldiers breathe before pulling a trigger. Those moments of stillness preserve civilization more than any innovation ever will. In a world obsessed with speed, slowing down is now a strategic act.

A data scientist once joked, "Our algorithms are getting so good, they might predict your mood before you do." A veteran beside him replied, "Then they'll know when to leave me alone." The room laughed, but the exchange carried weight. Technology may anticipate reaction, but it can't respect reflection. That's the difference between reading patterns and understanding people.

The great paradox of the 21st century is that the more we innovate, the more we need to remember what we already are. The human pulse, emotion, faith, intuition, and courage aren't outdated; they're endangered. As automation advances, rediscovering our humanity becomes not nostalgic but necessary. The nations that preserve it will not just lead the future, they will define it. Because the final competitive edge isn't speed, it's soul.

The human pulse is the metronome of civilization. It sets the pace, defines the rhythm, and reminds us when the music of progress risks turning into noise. Every sensor, system, and signal still relies on one thing it can't replicate: a person who cares enough to make it matter. When that care stops, the sky really does become empty.

In the next section, *"Faith as Framework: Seeing Stewardship, Not Domination, as the Highest Form of Innovation,"* we'll examine how belief, whether spiritual, ethical, or civic, gives direction to power. Because technology may provide us with wings, but only purpose decides where we fly.

381

Section 2 – Faith as Framework: Seeing Stewardship, Not Domination, as the Highest Form of Innovation

"We are not owners of progress; we are trustees of it." Thus, "Power without principle doesn't make you strong, it makes you dangerous.",
CSM Sheldon A. Watson.

We don't possess progress; we steward it. That's the line between civilization and conquest. Power, when divorced from principle, doesn't elevate; it endangers. And in an era where technology accelerates faster than wisdom matures, it is faith, not force, that anchors innovation to integrity. Without that anchor, advancement becomes intrusion, and brilliance becomes a breach.

Faith isn't confined to pews or prayers. At its core, it's conviction, a belief that life, work, and leadership serve something greater than self-interest. Every institution, from the family to the federal government, relies on faith in some form: faith in process, faith in promise, faith in people. Without it, trust collapses, and systems decay from the inside out. In this way, faith becomes the invisible scaffolding of civilization. It's what turns progress into stewardship, the belief that innovation should serve, not subdue, the world it touches.

The ancient temptation of humanity has always been domination, to rule over creation rather than care for it. In the age of automation, that temptation has better tools at its disposal. However, technology that dominates can become tyranny in disguise. We weren't meant to conquer the world through code; we were meant to cultivate it through conscience. Stewardship reframes innovation as guardianship, using ingenuity not to exploit, but to sustain. That's the kind of faith the modern world needs most: the belief that we are caretakers, not consumers, of the systems we build.

Innovation without ethics is a straight line with no destination. Faith, in any form, gives innovation shape, dimension, and boundaries. Like the laws of flight that keep a plane aloft, moral frameworks keep progress from stalling or crashing. Without lift and balance, even the best-designed machine fails. Faith provides that balance. It reminds us that the power to create must always remain tethered to the purpose of preserving. Progress asks, "Can we?" Faith asks, "Should we?" And the future depends on which question we answer first.

Restraint isn't a sign of weakness; it's a measure of wisdom. In leadership, in governance, and in innovation, restraint separates builders from opportunists. Faith enforces restraint by setting moral boundaries before technological ones. It keeps invention from outpacing intention. Without restraint, power becomes impatient, and impatience is the first step toward irresponsibility. The most advanced societies aren't the ones that innovate the fastest; they're the ones that innovate with foresight.

Hubris has always been humanity's blind spot, the belief that control equals competence. Every great fall in history, from empires to economies, began when confidence outgrew humility. Faith serves as a stabilizer, a quiet reminder that even the most innovative systems are fragile, and even the most advanced minds are fallible. It tempers ambition with awe. When leaders lose that sense of proportion, their organizations become efficient but soulless, productive but purposeless. Faith, in any form, puts the soul back into structure.

In today's world, stewardship takes many shapes:

- A programmer who builds transparency into code.
- A policymaker who pauses for public impact, not political gain.
- A commander who measures success not by destruction, but by discipline.

Stewardship is faith in motion, a belief that power can be principled and progress can be patient. It transforms innovation from consumption into conservation. The question is no longer who will lead the future, but who will safeguard it.

Every time we create something new, a policy, a platform, a drone, we also create responsibility. Faith transforms that responsibility into reverence. Reverence isn't fear of progress; it's respect for consequence. It's the humility to ask not just what technology can do, but what humanity can handle. That mindset turns invention into service and systems into sanctuaries.

Even in secular societies, faith in each other is what holds democracy together. Citizenship itself is an act of faith, belief in institutions, laws, and the unseen idea that unity is stronger than division. Automation can optimize governance, but it can't inspire it. Faith does that, the belief that leadership still matters, that service still means something, and that the country is worth more than the comfort of cynicism. Without that civic faith, even the most advanced democracy becomes mechanical, efficient, but hollow.

The best leaders operate on a form of faith: they believe in their people before proof, trust in their mission before victory, and are committed to principle before popularity. Faith-driven leadership doesn't require preaching; it requires practice. It shows up in patience under pressure, humility in success, and decency in defeat. That kind of faith builds institutions that outlast personalities because it sees power as a loan, not a possession.

During a conference on AI ethics, a developer joked, "We're teaching machines to learn faster than humans, let's hope they're also better listeners." A colonel replied, "If they start listening better than we do, that's not innovation, that's divine intervention." The laughter broke the tension, but the truth remained: listening, honest, patient, intentional listening, is still the purest act of faith.

Faith isn't about predicting the future; it's about preparing for it with humility. It guides innovation to serve humanity, not replace it. It tells us that control isn't the goal, contribution is. The difference between reckless ambition and righteous progress isn't intelligence; it's intention. And intention begins where faith resides: in the conviction that creation must always be matched with care.

Faith gives progress its coordinates. It reminds us that stewardship, not domination, is the true mark of civilization. The test of modern power isn't what we can automate, it's what we still choose to protect. Because in the end, every act of innovation carries a moral fingerprint. And every fingerprint should point upward, not inward.

In the next section, *"Home, Family, and Duty: Why Grounding Purpose in Relationships Sustains National Strength,"* we'll bring the discussion closer to earth, exploring how the strength of a nation doesn't begin in its labs or legislatures, but in its living rooms. Because technology may extend our reach, but only relationships extend our reason.

Section 3 – Home, Family, and Duty: Why Grounding Purpose in Relationships Sustains National Strength

"The strength of a nation begins in the strength of its homes." President Dwight D. Eisenhower.

"You can automate a system, but you can't automate sacrifice." CSM Sheldon A. Watson.

For two plus decades, I've watched aircraft lift off, precise, powerful, controlled. And every time, I'm reminded of the same truth: flight means nothing without a safe landing. The same applies to life, leadership, and progress. We can soar in innovation and strategy, but if we don't stay grounded in relationships, in duty, in love, in belonging, we risk becoming a nation that's airborne without purpose. Technology keeps us connected. Only people keep us committed.

Before any government, before any military, before any system, there was the family. It's the oldest and most reliable form of governance on Earth. Families teach the same principles that sustain democracies: responsibility, cooperation, fairness, and sacrifice. Strong families build strong citizens. And strong citizens build resilient nations. When families weaken, the fabric of civic life starts to tear, not because of policy, but because the habits of duty fade at home long before they fail in public.

Duty begins quietly, in how parents show up, how partners support each other, how communities stand together when life gets uncertain. It's not dramatic; it's deliberate. Duty is what bridges the personal and the national. It's the link between raising a child and defending a country, between helping a neighbor and holding a command. The same discipline that keeps a soldier alert in the field keeps a family firm in crisis: showing up, doing the work, even when no one's watching.

We live in an age of constant distraction. Devices promise connection but often deliver distance. We multitask conversations, delegate emotions, and measure attention in notifications. The tragedy isn't that technology consumes our time; it's that it consumes our focus. And focus is the foundation of care. When attention becomes fragmented, commitment becomes fragile. That's why grounding purpose in relationships is an act of resistance. It's how we reclaim depth in a shallow age.

Every home is a small republic, built on trust, accountability, and shared sacrifice. When those values erode inside households, they erode in institutions. Policy can't replace parenting. No government can legislate love or loyalty. Those have to be cultivated, meal by meal, conversation by conversation. A nation that forgets its families forgets its foundation. And a leader who neglects his own people stops understanding those he leads.

There's a kind of service that never makes headlines: the late-night talk with a child, the quiet forgiveness between spouses, the mentoring of a struggling teammate. That's the unpaid work that keeps freedom functioning. It's what turns citizens into stakeholders and followers into believers. The measure of a country isn't just its GDP or its drones, it's how many people still choose responsibility over convenience. Freedom isn't self-sustaining. It's home-sustained.

Every act of care reinforces the social armor that no adversary can breach. When we check on a neighbor, attend a community meeting, teach respect and patience to our children, we're doing national security work, even if it doesn't look like it. Because the most brutal enemy to defeat isn't external, it's apathy. And the surest defense against apathy is engagement at the most personal level.

The best leaders are rarely those who spend their time in their offices. They're the ones who understood people. They drew wisdom from the same place courage comes from, the home front. Family teaches what no doctrine can: empathy without compromise, firmness without cruelty, patience without passivity. It's where people learn to lead without authority and follow without resentment. Every command decision I've ever made drew, in some way, from lessons at home: listening before leading, correcting without humiliating, caring enough to hold standards. That's leadership born of love, not position.

The most important question for any generation isn't "What did we build?" It's "What did we teach?" When families instill purpose, the next generation doesn't just inherit wealth or education; they inherit ethos. They grow up understanding that responsibility is freedom's twin, not its enemy. A country that fails to transmit that lesson will soon find itself technologically rich but morally bankrupt.

During a family dinner, my youngest son once said, "Dad, you give orders at work, but here, Mom outranks you." He wasn't wrong. That's balance. And balance builds perspective. The same humility that keeps peace at the dinner table keeps perspective in command. Leadership begins at home, where you can't rely on titles, only trust.

When national conversations become toxic, it's often because personal conversations have disappeared. We debate loudly but listen rarely. We broadcast opinions but avoid accountability. Restoring civility at the national level begins with restoring it at the kitchen table. Because no algorithm can heal what only relationships can repair, technology can process conflict. Only humanity can resolve it.

The home is where principles are put into practice. Integrity is tested when no one is grading. Compassion is exercised when no one's posting. Family isn't just emotional support, it's moral rehearsal. It's where we learn how to handle power, navigate differences, and live with the consequences. Without that daily training ground, leadership loses its humanity.

Automation may change how we live. It can never change why we live. The purpose of progress is not to free us from responsibility; it's to give us more ways to honor it. The strength of a nation doesn't come from the altitude of its aircraft or the bandwidth of its networks. It comes from the steadiness of its families, the faithfulness of its people, and the quiet discipline of duty lived out daily. That's how civilizations endure, not through systems, but through souls.

In the next section, "The Reawakening: Citizenship as a Calling; Vigilance as Love of Country Made Visible," we'll expand from the home to the homeland, exploring how civic engagement is not merely a right, but a responsibility. Because the difference between a country that survives and one that thrives isn't power, it's participation.

Section 4 – The Reawakening: Citizenship as a Calling; Vigilance as Love of Country Made Visible

"The price of freedom is not comfort, it's commitment." Because "Patriotism isn't a feeling; it's a function. You prove it by showing up." CSM Sheldon A. Watson.

And in an age of automation, distraction, and division, showing up is no longer symbolic; it's strategic.

Every generation produces two kinds of citizens: participants and passengers. Participants carry the load. Passengers critique the journey. For too long, the majority has watched democracy like a spectator sport, cheering or jeering from the sidelines while a few do the heavy lifting.

But freedom doesn't thrive on observation; it thrives on ownership. The "silent majority" must reawaken, not as a political faction, but as a civic force. Because silence may keep peace for a season, but participation keeps liberty for a lifetime.

Citizenship was never meant to be passive. It's a vocation, one that requires attention, courage, and discipline. Voting, volunteering, mentoring, serving, these are not seasonal gestures; they're sustaining habits. Democracy doesn't die from attack; it dies from neglect. Citizenship is more than a passport; it's a pledge to take responsibility for the direction of the country, even when that direction feels uncertain. It means believing that one person's effort still matters, especially when cynicism says it doesn't.

There's a difference between vigilance and suspicion. Vigilance guards; suspicion divides. Vigilance means being alert to threats without assuming everyone's an enemy. It means asking hard questions, not harboring hard feelings. It's about protecting what's right, not policing what's different. Absolute vigilance is an act of love, love expressed through protection, not fear. It doesn't shout; it steadies. It doesn't accuse; it anchors.

Modern culture often confuses outrage with engagement. Outrage is reactive; it burns hot and fades fast. Engagement is disciplined; it sustains. Posting opinions isn't participation. Service is. Citizenship doesn't need more slogans; it needs more signatures: on volunteer rosters, voting rolls, and community boards. It's not about being loud, it's about being present.

Every national issue has a local address. Roads, schools, zoning, safety, public trust, these are the front lines of democracy. The most effective citizens aren't those who argue policy on screens; they're the ones who show up at the city council meeting, the PTA, the neighborhood watch. They know that protecting democracy begins with preserving the block. Because the national picture only changes when enough small pictures start to move.

Service today looks different from it did in uniform, but the spirit is the same. It's mentoring youth, helping veterans navigate systems, supporting first responders, cleaning parks, or teaching civic literacy. Every act of service is an act of stabilization. It says, I still believe this place is worth the effort. Service isn't about sacrifice anymore; it's about sustenance. It's how we keep the Republic from rusting.

Patriotism used to mean flags and parades. Today, it means showing up, not just on holidays, but on hard days. Filling a classroom seat, picking up litter, donating blood, mentoring a teenager, voting on a rainy Tuesday, those are acts of quiet patriotism. Love of country doesn't always roar. Sometimes, it whispers, I'll do my part.

Democracy is noisy by design. It's supposed to be. The challenge isn't the noise, it's the absence of reason inside it. Civil disagreement has become cultural warfare. But disagreement isn't a sign of division; it's proof of participation, when done with respect. Guarding the public square means defending both free speech and responsible speech. One without the other breeds chaos.

In a healthy democracy, leadership isn't limited to titles; it's distributed across temperament. The small-town coach who mentors kids away from trouble is as vital to the national character as any senator. The nurse who stays after shift because "it's the right thing to do" serves the Republic more than a thousand times. Leadership doesn't always wear rank; sometimes it just wears reliability. Every citizen can lead, if they remember that the smallest example can echo the longest.

Every privilege we enjoy, free press, free worship, free assembly, was once defended by someone who refused to stay silent. When citizens stop participating, institutions decay into self-preservation machines. Government becomes an observer, not a servant. And freedom begins to feel like furniture, taken for granted until it's gone. The enemy of liberty isn't tyranny, it's apathy. Because apathy doesn't break things, it just lets them fall apart.

The most powerful form of vigilance isn't suspicion, it's hope sustained by effort. It's believing that the country can still get better, and acting as if that belief depends on you. Hope doesn't mean naivety; it means endurance. It's the ability to see the cracks and still choose to build. That's the kind of patriotism the modern world needs: grounded, practical, relentless.

During a citizenship workshop, someone asked, "What's the quickest way to make a difference?" The instructor said, "Show up twice." Everyone laughed, but they understood. Consistency beats intensity every time. Change doesn't come from a single march or speech; it comes from habit.

Freedom is not self-cleaning. It requires maintenance, inspection, and, at times, renovation. Citizenship is that maintenance, the civic discipline of staying informed, staying civil, and staying engaged. It's the daily grind behind the grand ideal. Democracy isn't perfect, but it's repairable, as long as people still care enough to hold the tools.

Citizenship is not a gift; it's a duty disguised as one. And vigilance isn't paranoia, it's love made visible. The future of the Republic won't be decided by enemies abroad, but by engagement at home. Because the real test of patriotism isn't how loud you cheer when things go right, it's how steadfast you remain when they don't.

In the final section, *"Final Reflection: The Sky Isn't Empty – A Closing Meditation Tying the Title Back to Hope, Responsibility, and Shared Destiny,"* we'll bring the book full circle, returning to the title's meaning and leaving readers with one lasting message: that the sky above us is a mirror, and what fills it next depends on the choices we make together. Because the sky isn't empty, it's expectant.

Section 5 – Final Reflection: The Sky Isn't Empty – A Closing Meditation Tying the Title Back to Hope, Responsibility, and Shared Destiny

The future is not waiting to be discovered. It's waiting to be decided. And although the sky isn't empty, it's a mirror, showing exactly who we are and who we still have time to become." CSM Sheldon A. Watson.

Look up.

Every time you see the sky, you're not just seeing space; you're seeing a reflection. It holds the hum of drones, the streak of satellites, the contrails of aircraft, and the silence of possibility. The sky doesn't judge us; it records us. It's where our ambition lives and where our intentions collide. That's why this book began there. No frontier reveals human character faster than the one just above our heads.

The sky is not a void. It's a verdict.

We've built machines that think, systems that predict, and networks that never sleep. But the more efficient we've become, the more we risk losing the ineffable, the human touch that gives progress its meaning. Automation has redefined work, defense, and communication. But it hasn't replaced wonder, faith, or duty. The challenge of this century isn't whether we can innovate. It's whether we can integrate, whether we can advance without amputating the very things that make us human.

Today, our tools move faster than our ethics. Our reach exceeds our reflection. And yet, the future still bends to our will, if we dare to guide it.

The sky belongs to no one and everyone. Every drone that flies, every signal that transmits, every innovation that launches shares the same space. That makes airspace not just a domain, but a covenant. It's a reminder that progress without cooperation leads to collision. That no nation, no company, no citizen is exempt from responsibility in a world this connected.

Every launch is a promise. Every flight is a statement. Every silence between them is a question: What kind of civilization are we becoming?

Technology has given humanity a tremendous lift. But lift without direction is drift. And drift, whether moral, civic, or strategic, is how empires fall and societies forget themselves. That's why leadership, at every level, still matters. The pilot's discipline. The policymaker's judgment. The parents' guidance. The citizens' vigilance. All are part of the same equation: human responsibility steering mechanical power.

The machine can't care. The algorithm can't choose. But we can. And that choice defines the horizon.

Every generation inherits the sky in a different condition. Some inherit it wide open. Others inherit it, crowded with consequences. But all share one duty: to leave it safer, more transparent, and more navigable for those who follow. That's what stewardship really means, not preservation through paralysis, but protection through purpose. We aren't the first to shape the heavens. But we may be the last to do so carelessly.

Innovation will continue. The question is whether stewardship will keep pace.

Not every citizen will design a system or lead a mission. However, every citizen can contribute to the collective discipline that maintains the stability of the sky and society. The neighbor who reports suspicious activity. The teacher who teaches civic literacy. The developer who codes responsibly. These are the quiet defenders of modern freedom.

The future won't just be built in laboratories or launched from runways. It will be secured in living rooms, classrooms, and communities, where vigilance becomes a cultural norm. That's how ordinary people build extraordinary continuity.

At a technology symposium, a young engineer said, "Sir, I think AI will make humans obsolete." The old commander smiled. "Then who's going to teach it what matters?" The room went quiet. Because no one had an algorithmic answer for that, progress will always need perspective. And perspective still requires people.

Hope isn't naive; it's necessary. It's the only renewable energy that powers democracies through doubt. When people believe they can shape the future, they invest in it. When they stop believing, even the best technology can't save the system.

Hope is not a feeling; it's a discipline. It's the daily choice to engage when detachment feels easier, to build when cynicism feels safer. The sky reflects that discipline. When filled with reckless ambition, it becomes a threat. When filled with shared purpose, it becomes a possibility.

Every person carries a horizon inside them, a mental compass that defines how far they'll see and how much they'll care. Societies rise or fall based on whether that inner horizon expands or contracts. If we fill our inner sky with suspicion, fear, and apathy, the nation darkens. If we fill it with discipline, decency, and direction, the nation brightens.

The horizon within determines the one above. That's the secret of sustainable leadership, not commanding systems, but inspiring souls.

What happens next isn't predetermined. Technology doesn't dictate destiny; leadership does. Every drone, every law, every innovation is still subject to one force that can't be coded: the human conscience. That's our built-in autopilot, the quiet moral stabilizer that keeps the nose up when the weather turns.

If we maintain that, the future stays navigable. If we lose it, even the clearest sky becomes a battleground.

We began with a question: Who owns the sky? We end with an answer: No one. But everyone is responsible for it.

The sky isn't empty because it's filled with meaning, with the echoes of our ambition, the signatures of our choices, and the shadows of our stewardship. What we send into it, drones, data, dreams, reflects who we are. And what we protect within it reflects who we hope to become.

The horizon ahead is not fate, it's feedback. And it's still in our hands.

So look up, not for answers, but for accountability, because the sky isn't empty. It's waiting for direction.

Epilogue – The Quiet Sky

"The sky does not shout. It listens. And what it hears, it remembers."
"Vigilance isn't the noise we make; it's the quiet we maintain when the world forgets to look up.", CSM Sheldon A. Watson.

The Descent

Every mission ends the same way: descent. Engines throttle down, the ground reappears, and silence takes back what the noise borrowed.

That's how awareness works, too. It begins loud, urgent, full of data and direction, and ends in reflection. Because the real test isn't how high we climb in knowledge, but how calmly we land in understanding.

When I look at the sky today, I no longer see clouds or contrails. I see patterns, movement with meaning, silence with consequences. That awareness changes everything. Once you've learned to notice, you can't go back to looking without seeing.

The New Horizon

We live in an era when the sky has become a second internet, teeming with signals, data, surveillance, and autonomy. And yet, most people walk beneath it unaware.

That's not condemnation; it's a condition. Life is busy, and vigilance takes energy. But so does freedom.

401

The horizon ahead will not be defined by nations alone, but by networks, human, digital, and moral. Those who understand that balance will steer the next century. Those who ignore it will drift into it unprepared.

The Measure of Awareness

I've spent my life studying threats, tanks, rockets, improvised devices, and now, drones. But the most enduring threat has never been a weapon. It's been complacency.

Complacency whispers, *"That's someone else's job."* Awareness answers, *"No, it's ours."*

Every American, whether uniformed or civilian, is part of homeland defense.
Not by carrying arms, but by carrying attention.

The Hum Above

There's a sound now, faint but constant, the low hum of drones over neighborhoods, ports, and borders. To some, it's a nuisance. To others, it's progress. To me, it's a reminder.

That hum is the sound of a civilization in transition, learning how to balance innovation with responsibility. It's the heartbeat of a new era, one that will either protect us or expose us, depending on how disciplined we remain.

The hum above is not a threat or a comfort. It's feedback.

Quiet Doesn't Mean Safe

There's a myth that silence means peace. It doesn't. Silence is simply the space where preparation or neglect takes shape.

That's what "The Sky Isn't Empty" means. It's not a warning, it's a matter of awareness. The sky has always been full of movement, meaning, and moral consequence.

The difference now is that what fills it can reach everyone, everywhere, faster than ever before.

If that truth feels unsettling, good. Awareness should unsettle; it's how civilization stays awake.

The Weight of Vigilance

Vigilance is a burden that can't be automated. It has to be carried by people who still care enough to notice.

Every generation inherits the same question: *How much attention are we willing to give to what matters most?* That's not paranoia, that's patriotism in its purest form.

Because when vigilance fades, so does security. Not immediately, but inevitably.

The Next Watch

I often think about the young service members I've trained, their energy, their curiosity, and their commitment to getting it right. They don't just guard perimeters; they guard principles.

That's the real mission now: teaching vigilance that outlasts duty stations, uniforms, and eras. Turning awareness into culture, not just protocol.

The sky may be filled with drones, but it's also filled with duty, invisible, constant, shared.

A Moment of Humor

During a flight once, a passenger beside me looked out the window and said, "I wonder who's watching us from up there." I replied, "Hopefully someone who remembers they're being watched, too."

He laughed. Then he got quiet. Awareness does that, it doesn't paralyze; it provokes.

It reminds you that accountability goes both ways, from sky to ground and back again.

What Endures

Systems change. Tactics evolve. Technology upgrades. However, one thing that never loses relevance is character.

Character is what steadies the hand when tension rises, and what limits power when temptation whispers. You can't code it, clone it, or command it. You cultivate it.

That's the part of the mission that never ends.

The Quiet Sky

When I stand outside at night, I no longer see the sky as distant. I see it as a connection. Every satellite, every drone, every aircraft up there represents a choice made here on Earth. And that means the sky isn't separate from us; it's a mirror of us.

The quiet sky doesn't ask for applause or attention. It simply reflects our readiness to defend, to discern, to do better.

So look up, not with fear, but with focus. Because the sky isn't empty. It's waiting for direction.

Acknowledgments

Writing *The Sky Isn't Empty: America's Blind Spot in the Age of Drones* was never just a literary effort. It was a mission, one built on the shared experiences, mentorship, and sacrifice of countless people who shaped my journey both in uniform and in academia. To each of them, I owe more than words can measure.

To My Family

First and always, to my wife, Soleil, whose faith, patience, and brilliance have been my constant compass. You've weathered every deployment, every late-night study session, and every new chapter of this life with grace that humbles me daily.

My sons, Silas and Shane, are the next generation of leadership in my life. You remind me why I do this: to leave behind a world safer, smarter, and stronger than the one I inherited. You've taught me that legacy isn't about medals or titles; it's about presence, purpose, and perseverance.

You three are the quiet strength behind every sentence on these pages.

To My Mentors and Professors

To the faculty and staff at Liberty University's Helms School of Government, your guidance throughout my doctoral journey in Public Administration sharpened the analytical lens that inspired this book.

Your insistence on integrity in research and humility in leadership has stayed with me far beyond the classroom.

To my methodologist and committee reviewers who challenged every assumption and refined every argument, thank you for holding me to the same standard I have my soldiers: excellence without exception.

To the Profession of Arms

To the men and women of the United States Army, particularly those of NORAD & USNORTHCOM, and to every soldier, airman, sailor, marine, and guardian who has stood a post when the rest of the world slept, this book is as much yours as mine.

I've been privileged to serve among the best this nation has to offer: warriors who blend discipline with compassion, precision with purpose. Every insight in these pages, every lesson on vigilance, leadership, and civic duty, was forged in the crucible of your professionalism.

Special acknowledgment to my brothers and sisters at Task Force 51, ARNORTH 5th ARMY, the 1-9th Battalion, 1st Cavalry, and every team that made homeland defense not just a mission, but a movement. Your commitment to readiness shaped my understanding of what absolute protection looks like, not walls and weapons, but people who care enough to act.

To the Scholars and Innovators

To the thinkers, scientists, and engineers advancing drone technology, cyber defense, and artificial intelligence, your work embodies both the challenges and the promises of our time. This book would not exist without your brilliance and your courage to ask the hard questions about responsibility in innovation.

To my peers in the counter-Unmanned Aircraft Systems (c-UAS) community, from defense contractors to policy analysts, thank you for your relentless pursuit of solutions that protect freedom while honoring restraint.

We don't simply counter threats; we counter complacency. And that fight is constant.

To the Civil Servants and Citizens

To every law enforcement officer, emergency responder, municipal leader, and civic volunteer who keeps their communities safe with limited resources and unlimited heart, your quiet service is the foundation of national security. You remind the nation that vigilance isn't just a federal mandate; it's a local responsibility.

And to the American public, whose curiosity, courage, and conscience remain the strongest shield of democracy, this book was written for you.

You don't need a uniform to defend your country. You need awareness, empathy, and the will to participate.

Final Gratitude

Every page of this book bears the fingerprints of those who believed that leadership is a service to others and that knowledge should serve them.

Thank you to my editors, reviewers, and early readers who pushed for clarity without compromising conviction. Thank you to the soldiers who became scholars, and the scholars who became advocates. And thank you to the unseen hands who keep the sky, and our conscience, clear.

This book closes with my name on the cover, but it carries your echoes in every line.

To all who continue to serve, in uniform, in government, in community, may your vigilance never waver, and may your hearts never harden. Because the sky isn't empty, it's filled with your work.

Author's Note

When I began writing *The Sky Isn't Empty*, I didn't approach it as an author chasing publication; I approached it as a soldier who had spent years watching the world change from the ground up and the air down. For over two decades, I operated in a profession that demanded precision, preparation, and an unflinching awareness of the unseen. Yet nothing in my training quite prepared me for what I saw emerging above us: a new battlespace, one invisible to most and underestimated by many, the sky itself. Drones didn't arrive with fanfare or policy briefings; they crept in as tools, morphed into toys, and matured into tactics. In that evolution, I witnessed both the brilliance of innovation and the blind spots of complacency. Consequently, what began as a professional study, something I might have briefed to senior leaders at NORAD or USNORTHCOM, evolved into something far more urgent: a call to awareness for every citizen, policymaker, parent, and neighbor who still believes that security is not someone else's job.

Because the truth, stripped of jargon and ceremony, remains simple: defense begins at the ground level and ends only when we stop looking up. I wrote this book not to sound alarms, but to illuminate a gap, one that widens every time technology outpaces ethics and innovation sprints ahead while regulation limps behind. I've seen firsthand how awareness, when delayed, doesn't just lag; it invites danger to settle in quietly, without announcement or resistance. Drones, artificial intelligence, and automation do not inherently pose a threat to us. They

410

serve as mirrors, reflecting both our creativity and our carelessness. What we build reveals what we believe. If we build without purpose, we invite chaos. If we innovate without restraint, we invite regret. But if we lead with awareness, balance, discipline, and civic courage, we invite progress that serves humanity rather than replacing it. This isn't a book about fear; it's a book about focus. And if that sounds less dramatic than a dystopian thriller, good. The real threat isn't cinematic, it's systemic.

Throughout my years in uniform, I watched machines outperform men in speed, strength, and efficiency. However, I never saw one outthink a conscience. Every operation, regardless of its technological sophistication, still hinges on judgment, a human choosing to act with or without integrity. That's the part of leadership no algorithm can replicate. When I speak about "the sky," I'm not referring solely to altitude; I'm referring to perspective. Because what happens in the air remains as disciplined as what happens on the ground. And while the ground may be crowded with policy, politics, and public opinion, the sky remains a canvas, one that reflects our ambition, our accountability, or our absence of both.

When I entered Liberty University's doctoral program in Public Administration, I didn't do it to collect credentials like merit badges. I did it to understand the machinery of governance with the same depth I once studied the machinery of war. My capstone on counter–Unmanned Aircraft Systems (counter-UAS) policy exposed the tangled intersection of technology, law, and public trust. But what kept me up at night wasn't the complexity of the data; it was the realization that public awareness, the very foundation these systems were meant to protect, had grown dangerously thin. That gap between understanding and accountability didn't just concern me; it compelled me. It became the driving force behind this project.

Every statistic in this book points to a more profound truth: progress without partnership is fragility disguised as strength. Our ability to defend, deter, and adapt depends not just on funding or firepower, but on education, civic engagement, moral clarity, and technological literacy. The United States doesn't suffer from a shortage of capability; it suffers from a shortage of comprehension. And comprehension, unlike capability, cannot be outsourced. It requires citizens who see the whole field, who understand that awareness is not a luxury, it's a prerequisite. For this reason, this book was never meant to sit on an academic shelf gathering dust between dissertations. It belongs on kitchen tables, in command posts, and in the hands of anyone who's ever looked up and wondered who's watching the sky, and who's watching over it.

As I approach the close of my military career, I've come to understand that leadership is less about command and more about continuity. It's about ensuring the next generation inherits not just authority, but clarity, the kind that enables them to lead with both confidence and conscience. That's why I wrote *The Sky Isn't Empty* not as a warning, but as a passing of the torch. Awareness doesn't require a title or a clearance level; it requires a posture, one that every citizen can adopt, regardless of background or expertise. If I've learned anything from my time in armor, command, and strategy, it's this: the best defense is an educated people. And education, unlike equipment, doesn't expire or malfunction. It scales. It sustains. It empowers.

At the core of this book lies a simple conviction: we can no longer afford to separate the conversation about technology from that about humanity. What we fly, we must also guide. What we automate, we must also answer for. And what we invent, we must also understand. The sky above us serves as both stage and symbol. It reflects our ingenuity, our ambition, and our accountability, or our absence of it. I hope that this work inspires readers to think differently, to look upward with both curiosity and caution, and to remember that freedom, like flight, depends on lift and balance. Because progress without principle isn't progress at all, it's drift. And drift, as history reminds us, is how civilizations lose altitude.

This book doesn't center on drones; it centers on the distance between awareness and action, and what happens when that distance widens. If even one reader walks away seeing the sky not as a backdrop but as a shared responsibility, if one leader reconsiders how they define vigilance, then this effort has done its job. The sky isn't empty. It waits for direction. And it waits for all of us to take the controls, not with fear, but with focus; not with panic, but with purpose.

413

The Reader's Challenge

"The duty of a free people is not to wait for orders, it's to recognize them when history issues the call.", U.S. Defense Forum, 2024. "Awareness without action is just observation. And observation never won a fight." CSM Sheldon A. Watson.

The Hand-Off

If you've made it this far, you already understand that this book isn't a conclusion; it's a continuation. Every chapter was designed to expose one truth: that the defense of a nation doesn't begin in boardrooms, command centers, or laboratories. It starts with awareness, with citizens who remain attentive, care, and take action.

That responsibility now belongs to you.

The tools of vigilance are no longer limited to uniforms or clearance badges. They're accessible to anyone with curiosity, courage, and a conscience.
The challenge is simple: don't let comfort turn into complacency.

Because in a world where threats move at the speed of technology, inaction is a luxury we can't afford.

The First Step Is Attention

Awareness is not paranoia; it's preparation. The difference between the two is perspective. Paranoia isolates. Preparation integrates.

414

Every community, every family, every agency needs people willing to observe, report, and respond with discipline rather than panic.

It's easy to assume someone else is watching. But as I've said throughout these pages, *the sky belongs to all of us.* That means responsibility does, too.

When you hear the hum of a drone, see strange behavior around critical sites, or notice gaps in policy or communication, don't dismiss them. Document. Report. Educate.

In today's world, the bystander is as dangerous as the adversary.

The Power of Participation

Democracy doesn't need more opinion; it needs more participation. Be the one who shows up at a council meeting, supports local emergency training, or joins a volunteer group focused on public safety and technology awareness.

If you're in business, advocate for ethical innovation. If you're in government, insist on interagency communication and accountability.

If you're in education, teach students not just how to use drones, but how to use judgment.

And if you're just a concerned citizen, remember: citizenship is an action verb. You don't have to change the world. Just change the five feet around you, consistently, deliberately, and visibly.

Bridging the Civil-Military Divide

We talk often about the "gap" between the military and civilian worlds. In truth, that gap isn't cultural, it's conversational.

Service members learn vigilance by necessity. Civilians must know it by choice.

Bridging that gap requires dialogue, respect, and humility, with soldiers sharing lessons and civilians sharing perspectives, both realizing that national defense is a shared language.

When those two worlds start speaking again, the country remembers its rhythm.

Leadership Without Rank

You don't need a title to lead. Leadership begins wherever you decide to take responsibility.

That could mean mentoring youth, starting a neighborhood safety initiative, or simply being the person in your workplace who still believes ethics matter. Authentic leadership isn't about control; it's about consistency.

Technology has made leadership scalable. A single act of integrity, shared at the right time, can ripple across a nation faster than a broadcast.

That's how cultural course corrections begin, not in institutions, but in individuals who act with conviction.

Hope as a Discipline

You'll hear people say, "Things are getting worse." That's not leadership, that's surrender dressed as commentary.

Hope isn't unquestioning optimism; it's disciplined endurance. It's the act of building even when the system appears to be broken. It's the decision to keep faith in people even when politics disappoints.

Cynics never wrote the history of turning points. They were written by those who stayed at their post, steady, ready, and stubbornly hopeful.

Choose hope. It's not soft. It's strategic.

From Awareness to Action

Now that you've seen the whole picture, the risks, the realities, the responsibilities, the next move is yours. You can't unknow what you know, and that's the point.

Use it.

Teach it.

Live it.

Start conversations about drone awareness in your community. Push for balanced legislation that encourages innovation but enforces accountability.

Support local responders with resources and recognition. And above all, keep your eyes on the sky, not in fear, but in awareness.

Because awareness, when shared, becomes culture. And culture, when disciplined, becomes defense.

A Moment of Humor

A colleague once told me, "You can't expect people to stay alert forever."

I smiled and said, "That's why we call it a watch, not a nap."

The joke got a laugh, but the meaning stuck. We don't stay alert because we enjoy stress. We stay alert because freedom demands endurance.

That's the cost of living in a free society, staying awake long enough to preserve it.

Your Mission

So, here's the challenge: Before you put this book on a shelf, decide what you'll do with what you've read.

Will you volunteer?

Will you educate your children about digital responsibility? Will you advocate for more innovative policies in your community? Or will you look up a little more often and notice what others ignore?

Every great movement starts with a simple shift, from *awareness* to *ownership*.

That's your mission now.

No uniform required.

The Last Word

I've spent my life serving a flag that represents something rare, a country that still trusts its people to be part of its defense. That trust is earned, not assumed.

So, when you look at the sky, whether filled with clouds, drones, or stars, remember: it's a reflection of what we allow, what we build, and what we protect.

The sky isn't waiting for Washington. It's waiting for us.

This is your watch now. Keep it steady. Keep it human. Keep it hopeful.

Because the sky isn't empty, and neither is your responsibility.

Appendix A: The Shared Responsibility Doctrine

"Security is not a spectator sport; it's a shared command.", CSM Sheldon A. Watson

1. Purpose and Rationale

The Shared Responsibility Doctrine (SRD) defines homeland resilience not as a government program but as a civic covenant. It recognizes that no single agency, algorithm, or uniform can safeguard a nation whose people remain unaware of the sky above them.

This doctrine challenges three assumptions that weaken modern security culture:

That safety is someone else's job.

That awareness equals fear.

That technology can replace judgment.

The SRD reframes national defense as a three-tier partnership, citizens, institutions, and governments, bound by trust, informed awareness, and disciplined communication. Its mission is simple: turn spectators into sentinels and responders into collaborators.

2. The Three Levels of Readiness Model

Level I: Individual Awareness

Citizens are the first line of defense in any threat-detection network. Their vigilance is neither paranoia nor policing; it is participation.

- **Recognize Patterns:** Note unusual drone activity, electronic interference, or sudden signal loss near public spaces.

- **Record Responsibly:** Capture time, location, and description; never interfere or engage.

- **Report Promptly:** Local law enforcement → State Fusion Center → FAA Hotline (1-844-FLY-SAFE).

- **Respect Boundaries:** Awareness ends where aggression begins; shooting or jamming drones is illegal under federal law.

A vigilant citizen observes without escalating. Awareness replaces anxiety with agency.

Level II: Institutional Responsibility

Organizations, such as schools, businesses, faith communities, and corporations, form the middle ring of national resilience. Their duty is twofold: to *educate their members* and *maintain ethical control over the technologies they employ.* Key obligations include:

Establishing standard operating procedures for drone incidents or airspace violations.

421

Conducting annual training or tabletop exercises simulating unauthorized drone events.

Appointing a Technology Ethics Officer or safety liaison to ensure compliance with FAA and local policy.

Maintaining clear communication channels with public-safety agencies.

Institutions that wait for a crisis to reveal their weaknesses have already failed the readiness test.

Level III: Intergovernmental Coordination

Governments, federal, state, and local, form the outer ring of resilience. Their mandate is to synchronize intelligence, regulation, and rapid response while maintaining transparency and accountability. This requires:

- **Data Interoperability:** Unified reporting platforms linking FAA, DHS, DoD, and state emergency networks.
- **Joint Training Exercises:** Annual "Blue Sky Drills" combining civilian responders, Guard units, and local agencies.
- **Policy Agility:** Laws updated at the speed of technology, not the pace of bureaucracy.
- **Public Transparency:** Quarterly community briefings on drone incidents and mitigation measures.

Intergovernmental coordination succeeds when each level can operate independently while speaking the same operational language during a crisis.

3. Doctrine in Action

Scenario 1: Unauthorized Drone Over a Public Event

A drone appears over a city marathon.

Level I: Spectators record and submit footage via the local CRC app.

Level II: Event security activates its pre-established SOP and contacts the FAA liaison.

Level III: Local police coordinate with NORAD/NORTHCOM liaison through the CISA portal. Result: Rapid identification, controlled response, zero panic.

Scenario 2: Infrastructure Interference

A utility company detects radio interference near a substation.

Level I: Employees note and log timestamps.

Level II: Corporate security cross-checks flight data with FAA records.

Level III: The state fusion center initiates the counter-UAS protocol. Outcome: Threat neutralized before impact.

Each scenario demonstrates the doctrine's principle: awareness without coordination is noise; coordination without awareness is blind.

4. Leadership and Ethical Mandate

The SRD is not a legal requirement; it's a moral one. Leadership at every level must cultivate a culture where vigilance is regular, communication is disciplined, and ego never outranks ethics. Commanders, mayors, principals, and CEOs share the same responsibility to act before they are forced to react.

Guiding Tenets:

Transparency builds trust.

Preparation is proof of integrity.

Ethical clarity prevents operational chaos.

5. End State, Resilience as Culture

Success under the Shared Responsibility Doctrine will not be measured by the number of threats neutralized but by the speed, unity, and calm with which a community responds.

When a neighborhood report seamlessly integrates into a city dashboard, when a teacher knows whom to call before a parent panics, and when faith leaders can reassure their congregations with facts rather than fear, that is the mark of a nation whose people have reclaimed stewardship of their own security.

The doctrine's end state is simple yet profound:

A country where citizens do not wait to be rescued, they participate in their own defense.

Appendix B: Drone Incident Community Checklist

"Panic is contagious. So is composure." CSM Sheldon A. Watson.

Purpose

This checklist provides a simple, standardized response protocol for civilians, civic leaders, and local institutions who encounter unidentified or unauthorized drone activity. It ensures safety, upholds the law, and promotes clear communication among citizens, agencies, and authorities.

The goal is to replace fear with structure and confusion with coordination.

1. Immediate Actions: First 60 Seconds

If you observe a drone behaving unusually:

- **Stay Calm and Stationary.** Avoid drawing attention or confronting the operator.
- **Do Not Engage.** Do not throw objects, use firearms, or attempt to jam the signal. These actions are illegal under federal law and can endanger others.

- **Move People to Safety.** If the drone is low, erratic, or near crowds, quietly create distance and guide others away from the area.

- **Document, Don't Dramatize.** Record short, steady footage (using a phone camera with a timestamp visible). Avoid speculation or posting on social media during active observation.

- **Call It In.** Notify local law enforcement or event security immediately. Use clear, factual language.

Rule of thumb: When in doubt, treat every unidentified drone near people, power, or property as a potential safety risk, never a curiosity.

2. Observation Protocol: What to Note

Capture as many details as possible without approaching the drone or operator:

Category	Details to Record
Time & Date	Exact observation time (24-hour format preferred).
Location	GPS coordinates or street address; include landmarks.

427

Category	Details to Record
Altitude / Behavior	Estimate height, movement pattern (hovering, circling, descending).
Appearance	Color, size, number of propellers, attached payloads, or lights.
Sound	Loud/quiet; steady hum or pulsing buzz.
Direction of Flight	Compass direction or relative movement (toward school, highway, event, etc.).
Operator Sightings	Any visible individuals with controllers, vehicles, or antenna setups.
Photos / Videos	Capture 5–10 seconds of footage, keeping people's faces out when possible.

3. Reporting Chain: Who to Notify, and How

Follow this chain of escalation until acknowledged by a responsible authority:

Local Law Enforcement: Non-emergency dispatch (or 911 if immediate threat).

Provide a factual observation summary.

State Fusion Center / Emergency Management, when confirmed as suspicious or repeated.

Example: *Colorado Information Analysis Center (CIAC) / Texas Fusion Center.*

Federal Authorities

FAA Hotline: 1-844-FLY-SAFE (1-844-359-7233)

DHS CISA Reporting Portal: www.cisa.gov/report

FBI Field Office: Contact if the pattern suggests criminal or cross-state coordination.

Internal Institutional Reporting

A school safety officer, security contractor, or church safety lead should log the details in the local incident record and notify the leadership chain.

Communicate once, accurately. Avoid multiple redundant calls that flood local dispatch. Send evidence and notes to the first confirmed point of contact.

4. Legal Boundaries; What Not to Do

Prohibited Responses (Federal Offense):

Shooting, disabling, or attempting to seize the drone.

Using signal jammers, GPS spoofers, or microwave devices.

Entering restricted areas or trespassing while pursuing an operator.

Permitted Responses:

Filming or photographing in public spaces.

Notifying property owners or event coordinators.

Providing factual statements to authorities.

Federal law (18 U.S.C. § 32 & § 1367) criminalizes destruction or interference with aircraft, including drones. Violators can face fines or imprisonment.

5. Post-Incident Follow-Up

Once the situation stabilizes:

File an Internal Report: Time, place, participants, summary, attached photos/videos.

Request a Case Number: From local law enforcement for future reference.

Conduct an After-Action Reflection:

What worked?

What caused confusion?

What can be improved before the next event?

Share Lessons Learned: With your Civic Resilience Council (CRC), school, or community board.

Update Your SOP: Incorporate Improvements into the Next Safety Briefing.

Preparedness is iterative. Every incident becomes a rehearsal for the next.

6. Communication Discipline

When sharing information publicly:

Post **only verified details** cleared by authorities.

Avoid naming individuals or speculating about motives.

Use official hashtags or emergency-alert channels to avoid misinformation.

Remember: in the digital age, false reports spread faster than drones fly, and can cause greater damage.

7. Quick-Reference Summary

Step	Action	Objective
1	**Observe**	Stay calm; assess the situation.
2	**Record**	Capture details safely.
3	**Report**	Contact the proper chain.
4	**Reinforce**	Follow up, learn, and adapt.

End State

A resilient community isn't one that never faces threats; it responds with discipline, clarity, and calm coordination. Every accurate report, every responsible witness, and every informed leader strengthens the nation's collective situational awareness.

"Preparedness is patriotism practiced locally." CSM Sheldon A. Watson.

Appendix C: Civic Technology Education (CTE) Framework

"Teach them early that vigilance is not fear, it's respect in action.", CSM Sheldon A. Watson.

1. Purpose and Vision

The Civic Technology Education (CTE) framework equips future citizens to navigate a world in which the sky, the web, and the public square are interconnected domains of responsibility.

Its goal is to merge technological literacy with civic virtue, ensuring that curiosity never outruns conscience. Where traditional STEM programs teach *how* to innovate, CTE teaches *why* to innovate responsibly.

2. Guiding Principles

- **Humanity Before Hardware**: Innovation Must Serve Human Dignity, Not Convenience Alone.
- **Transparency as Trust**: Digital citizenship demands openness, accuracy, and respect for privacy.
- **Preparedness Over Panic**: Education Replaces Fear with Informed Confidence.
- **Faith and Ethics as Frameworks**: Moral reasoning belongs in every lesson on technology.

433

- **Community as Curriculum**: Learning is strongest when tied to real civic impact.

3. Curriculum Vision: "Teaching Tech Through Character"

Level	Learning Theme	Focus Areas	Example Activities
Elementary (Grades K-5)	*Curiosity & Responsibility*	Understanding what drones and smart tech do, and respecting others' privacy.	"Design Your Own Safety Sky" art project • Identify 'good use vs bad use' scenarios • Build paper drones to learn lift & ethics.
Middle School (Grades 6-8)	*Ethics & Awareness*	How Data, Cameras, and Connectivity Shape Society.	Classroom debate: "Is Technology Neutral?" • Drone-demo day with local police UAS unit • Short essay: "If I Had a Drone."

Level	Learning Theme	Focus Areas	Example Activities
High School (Grades 9-12)	*Civic Technology & Law*	FAA rules, privacy rights, digital forensics, civic duty.	Simulated city-council hearing on drone zoning • Guest lecture from FAA or CISA rep • Student-run awareness campaign.
Post-Secondary / Community	*Applied Responsibility*	Emerging tech policy, AI ethics, dual-use systems.	Capstone: "Design a Drone Policy for Your Town" • Joint faith-civic ethics roundtable.

4. Implementation Roadmap

Step 1: Integrate, Don't Add

Fold CTE modules into existing subjects, science, civics, technology, and ethics. *A physics lesson on flight can end with a question on moral flight paths.*

Step 2: Build Local Partnerships

Collaborate with:

Local FAA Safety Teams or airport outreach programs.

State Departments of Education & Emergency Management.

Veteran and Faith Organizations for mentorship talks.

Step 3: Train the Teachers

Provide short, certified workshops on drone safety, public policy, and digital ethics. Each educator should leave with three takeaways: *Know the law, teach the limits, and inspire the "why."*

Step 4: Engage Families

Send home a "Sky Safety Pledge" to encourage family discussion about responsible tech use.

5. Annual Event: "Sky Safety Week"

An annual, school-community initiative promoting awareness and unity in technology ethics.

Day	Theme	Activities
Day 1, Awareness Launch	Opening assembly: "Who Owns the Sky?" keynote • Student drone demonstration.	
Day 2, Policy in Practice	Mock city meeting on airspace regulations • Poster contest: "My Sky, My Responsibility."	
Day 3, Safety & Ethics	Emergency-response drill • Faith and community leaders panel: "Moral Courage in the Modern Age."	
Day 4, STEM Showcase	Students present tech projects to local officials and veterans • Coding challenge: "Predict & Prevent."	
Day 5, Community Commitment	"Blue Sky Oath" ceremony • Drone flyover for unity photo • Recognition of student ambassadors.	

6. Faith and Community Partnerships

Faith communities remain the moral infrastructure of resilience.

Host joint workshops with schools on ethics in innovation.

Train youth ministers and volunteer leaders as Faith Liaisons for Technology Awareness.

Use sermons and study groups to discuss stewardship, justice, and humility in the digital age.

"When morality keeps pace with machinery, civilization endures."

7. Assessment & Metrics

To sustain credibility, each participating school or organization should report annually on:

Number of CTE modules delivered.

Student engagement scores.

Incident response drills completed.

Partnerships forged with law enforcement or industry.

Metrics don't replace meaning, but they prove momentum.

8. End State: A Literate Nation

When a generation can name a drone model as easily as a moral consequence, when students debate policy before crisis forces policy upon them, and when parents discuss ethics with the same ease they discuss homework, that is civic literacy realized.

"We can't program virtue, but we can teach it." CSM Sheldon A. Watson.

Appendix D, Policy and Regulatory Quick Reference

"Law can't ride shotgun from the rearview mirror.", CSM Sheldon A. Watson.

1) FAA Core Framework (Civil U.S. Airspace)

Part 107, Small Unmanned Aircraft Rule (Commercial/Non-recreational)

Who: Most non-recreational operators.

Key requirements: Remote pilot certificate; aircraft <55 lb; registration; operate within visual line of sight (with limited waivers); altitude, airspace, and operating limitations; additional allowances for operations over people, over moving vehicles, and at night when conditions are met. FAA+1

Reference: FAA Part 107 overview + "Operations Over People" rule update. FAA+1

49 U.S.C. § 44809, Recreational Flyers (Statutory Exception)

Who: Hobby/recreational flyers only.

Key conditions (high level): Fly strictly for recreation; follow an FAA-recognized CBO safety code; pass TRUST; register (if required); give way to manned aircraft; fly in authorized airspace; Remote ID compliance (unless eligible exceptions). Legal Information Institute+1

Reference: U.S. Code text + FAA recreational page (updated 2025). Legal Information Institute+1

Remote ID, "Digital License Plate" for Drones

What: Broadcasts the identity and location of the drone and control station to authorized receivers; required for most operations, with Letters of Authorization available for specific research/waiver contexts. FAA

Why it matters to communities: Enables law enforcement and authorized agencies to correlate a flight with an operator, critical for stadiums, parades, and incident response. (Recent policy discussion emphasizes real-time access for SLTT agencies.) ucdrones.github.io

2) Defense & Security Authorities (Domestic C-UAS)

10 U.S.C. § 130i, Protection of Certain Facilities and Assets

What it authorizes (DoD at designated locations): Detect, identify, monitor, track, and, where authorized, mitigate UAS threats (including actions that would otherwise violate surveillance or anti-tampering laws), subject to strict scope and oversight. Legal Information Institute+1

Why communities care: DoD authority is site-bound and purpose-bound; it does not generalize to local police or private security.

Coordination with FAA/DHS remains essential. (CRS overview provides current posture and legislative activity.) Congress.gov

DHS/DoJ C-UAS Posture (Context)

DHS reports continued implementation of limited C-UAS authority for selected missions/facilities; expansion efforts remain under congressional review. (Useful context for interagency talks.) Senate Judiciary Committee

3) Interagency & Emerging Policy Notes

FAA Reauthorization (2024–2025): Ongoing updates emphasize integrating Remote ID with authorized real-time access for federal and SLTT agencies and clarifying the roles in detection and mitigation. Track local implementation guidance. ucdrones.github.io

Market reality check: Manufacturer geofencing is not the law; major vendors have shifted from hard "no-fly" blocks to warning-only models, placing more weight on operator compliance and Remote ID enforcement. (Operational risk for wildfire zones, airports, sensitive sites.) The Verge

Legal guardrails for everyone else: Outside narrowly defined federal authorities (e.g., § 130i sites), shooting down, jamming, or spoofing drones can violate federal criminal and communications law. Coordinate through FAA/LE channels rather than resorting to ad hoc countermeasures. (See Part 107/§ 44809 + DOJ/DHS guidance in local SOPs.) FAA+1

4) "At-a-Glance", Who Can Do What?

Scenario	Local Police / Venue Security	State Fusion Center	FAA	DoD (at § 130i sites)
Identify & track unknown drone.	Observe, record, and request Remote ID assist via authorized channels	Aggregate intel, coordinate partners	Airspace authority; Remote ID policy	Detect/ID/monitor within authorized perimeter
Mitigate (disable/interdict) the drone	No (generally	No (unless operating under federal	No (regulator; may coordinate	Yes, within strict statutory scope (§ 130i)

443

Scenario	Local Police / Venue Security	State Fusion Center	FAA	DoD (at § 130i sites)
	prohibited)	authority)	TFRs, NOTAMs)	
Set airspace rules for events	Request TFRs; enforce local ordinances consistent with federal law	Coordinate requests and threat intel	Issue TFRs/waivers; national airspace integration	N/A (unless mission overlaps protected asset)

(Your stadium, parade, wildfire, and critical-infrastructure vignettes fit squarely here: report → coordinate → request federal tools; don't freelance mitigation.)

5) Practical Links & Numbers (for Checklist/SOPs)

FAA Remote ID / Policy: faa.gov/uas/getting_started/remote_id (policy & LOA info). FAA

Recreational Flyers: faa.gov/uas/recreational_flyers (TRUST, CBO codes, airspace basics). FAA

Part 107 Overview: FAA newsroom/Part 107 + Ops Over People update. FAA+1

U.S. Code: 49 U.S.C. § 44809 (recreation); 10 U.S.C. § 130i (C-UAS authority). Legal Information Institute+1

CRS Brief on DoD C-UAS: congress.gov CRS R48477 (2025). Congress.gov

Report suspicious UAS: FAA Hotline 1-844-FLY-SAFE; coordinate via state fusion centers; CISA reporting portal, as per local SOPs. FAA

6) Plain-English Bottom Lines

Remote ID is the backbone of enforcement; it ties flights to operators. Expect growing real-time access for authorized agencies. FAA+1

Part 107 and § 44809 define your lane. Commercial vs. recreational rules differ, but both point back to FAA authority. FAA+1

Mitigation is not DIY. Outside narrowly defined federal perimeters, disabling a drone can be a federal crime. Coordinate, don't improvise. Legal Information Institute

Geofencing ≠ law. Don't assume the software will save you; plan communications, templates, and responses as if you'll need them. The Verge

7) Suggested Reading

Advisory Circular 91-57B (recreational ops guidance under § 44809). FAA

Chaffin, B., Gosnell, H., & Cosens, B. (2014). Adaptive Governance: From Theory to Practice.

DHS Strategic Plan FY23–27 (mentions C-UAS posture & integration).

FAA Part 107 PDF Summary (operational constraints & pilot duties). FAA

FAA. (2021). UAS Integration Pilot Program Summary.

Folke, C. (2016). Resilience (Republished). Ecology & Society.

GAO. (2023). Counter-Drone Technologies: DoD Coordination Challenges.

Lin, P., Abney, K., & Bekey, G. A. (2012). Robot Ethics: The Ethical and Social Implications of Robotics. MIT Press.

Singer, P. W. (2009). Wired for War. Penguin.

Appendix E, Ethical Leadership in the Age of Automation

"Technology tests capability; ethics tests command.", CSM Sheldon A. Watson

1. Purpose

This appendix equips leaders in, military, civil service, and corporate, to navigate the moral turbulence created when automation and artificial intelligence extend authority beyond human reach. Its mission is simple: preserve conscience in command.

2. The Moral Algorithm: Decision-Making Under Automation

Automation accelerates choices; ethics must accelerate discernment. The Moral Algorithm provides a disciplined, five-step process:

- **Clarify the Context.** Identify what the machine sees, and what it can't.
- **Confirm Authority.** Determine who owns the consequence. *If everyone owns it, no one does.*
- **Check for Bias.** Examine training data, assumptions, and unintended discrimination.
- **Challenge the Outcome.** Ask: *Does this serve mission and morality?*
- **Communicate the Decision.** Transparency transforms trust into policy.

If a decision cannot survive sunlight, it should not survive automation.

3. Command Without Command: Leadership Presence in Autonomous Systems

When machines execute faster than orders can be transmitted, leadership must shift from control to culture.

- **From Orders to Oversight:** Replace Step-by-Step Directions with Outcome-Based Intent.

- **From Silence to Signal:** Establish "ethical override" channels where any operator can question a system's action without reprisal.

- **From Metrics to Meaning:** Evaluate missions not only by efficiency but by integrity, the manner in which success is achieved.

Autonomy without accountability is abdication.

4. The Five Core Ethical Anchors

Anchor	Definition	Practical Expression
Stewardship	Treat technology as a trust, not a toy.	Use every system as if the public were watching.

Anchor	Definition	Practical Expression
Accountability	Own outcomes, even those delegated to code.	Sign your name to the algorithm's decision path.
Humility	Recognize that machine speed doesn't equal wisdom.	Pause before approving irreversible automation.
Empathy	Understand human impact before data impact.	Ask who benefits, who's burdened, who's unseen.
Transparency	Keep ethics visible.	Publish decision logs, not excuses.

5. Case Study 1, The Autonomous Decision Error

- **Scenario:** A law-enforcement drone identifies a "suspect" vehicle via outdated facial-recognition data and alerts a patrol to intercept. The vehicle belongs to a city councilwoman returning from a veterans' fundraiser.

- **Failure Point:** Automation was correct in process, wrong in principle; the system acted faster than oversight.

- **Ethical Correction:** Immediate suspension of auto-dispatch protocols.

- Establishment of a *Human-in-the-Loop* review layer.

- Public apology from agency leadership, reinforcing accountability over accuracy.
- **Lesson:** *Speed does not sanctify error; owning it restores credibility.*

6. Case Study 2: The Faith-Based Intervention

Scenario: During a humanitarian mission, an autonomous targeting system designates a suspected weapons cache inside a church-run medical clinic. A field commander halts the strike despite confirmation from the algorithm.

Outcome: Subsequent ground verification confirms the commander's correctness. The "cache" was medical equipment wrapped in foil.

Ethical Reflection: The commander's moral pause saved lives and preserved legitimacy. Faith and reason cooperated where automation could not.

Lesson: *When conscience intervenes, technology learns humility.*

7. Reflection Questions for Leaders

Does my team know who holds final moral authority in automated operations?

Do we reward speed over wisdom?

Are our systems trained in truth or convenience?

Have we rehearsed "stop-order drills" where ethics take precedence over execution?

Could my decision be briefed publicly without shame or secrecy?

Leadership isn't about keeping pace with technology; it's about maintaining purpose ahead of it.

8. End State; Moral Readiness

Actual readiness isn't just technical; it's ethical. A leader's credibility is the last line of defense when code misfires and chaos follows. The nation that guards its conscience as fiercely as its perimeter will never be outflanked by its own machines.

"Integrity must travel at the same speed as innovation.", CSM Sheldon A. Watson.

Appendix F: Glossary and Acronyms

"Clarity is the first act of leadership.", CSM Sheldon A. Watson.

Term / Acronym	Definition
AI (Artificial Intelligence)	Computer systems that are capable of performing tasks that typically require human intelligence, such as perception, decision-making, and pattern recognition.
AO (Area of Operations)	The geographic region in which assigned military or interagency missions are executed.
AUTONOMY	The ability of a system, particularly a drone or algorithm, to perform tasks without direct human control.
BLOS / BVLOS (Beyond Visual Line of Sight)	Drone operations conducted beyond the pilot's unaided visual range require an FAA waiver and advanced detection systems.

Term / Acronym	Definition
C2 (Command and Control)	The authority and processes through which a commander directs and coordinates forces and operations.
C4ISR	Command, Control, Communications, Computers, Intelligence, Surveillance, and Reconnaissance, core functions integrating technology and decision support in defense operations.
CISA (Cybersecurity and Infrastructure Security Agency)	A division of the U.S. Department of Homeland Security responsible for protecting critical infrastructure and coordinating national cybersecurity.
CIVILIAN RESILIENCE COUNCIL (CRC)	A proposed community-based structure, introduced in The Sky Isn't Empty, aims to coordinate public awareness, preparedness, and communication between citizens and authorities.

Term / Acronym	Definition
COA (Course of Action)	A proposed operational plan or method for achieving an objective within set constraints.
CONOPS (Concept of Operations)	A narrative or graphical statement describing how a system or organization operates to achieve mission goals.
C-UAS / C-UAS (Counter–Unmanned Aircraft System)	The detection, tracking, identification, and mitigation of unauthorized or hostile drones.
DHS (Department of Homeland Security)	U.S. federal department responsible for protecting the nation against threats, including terrorism, cyberattacks, and natural disasters.
DoD (Department of Defense)	The federal department tasked with coordinating and supervising all agencies and functions of the government directly related to national defense.

Term / Acronym	Definition
FAA (Federal Aviation Administration)	The agency within the U.S. Department of Transportation responsible for regulating civil aviation and national airspace safety.
FCC (Federal Communications Commission)	Regulates communications across radio, television, wire, satellite, and cable; manages the electromagnetic spectrum used by drones and wireless systems.
GAO (Government Accountability Office)	The legislative branch agency that audits and evaluates U.S. government programs, including defense and homeland security systems.
GEO-FENCING	Software-based boundary limiting a drone's movement within predefined airspace. Violations trigger alerts or automatic restrictions.
HUMAN-IN-THE-LOOP (HITL)	A system design requiring human input or supervision during critical stages of automated decision-making.

Term / Acronym	Definition
IC: (Intelligence Community)	The collective group of 18 organizations that gather and analyze national security information for the United States.
ISR (Intelligence, Surveillance, Reconnaissance)	Integrated processes that collect, process, and disseminate information to support decision-making.
JADC2 (Joint All-Domain Command and Control)	The U.S. military framework for connecting sensors, shooters, and decision-makers across domains (air, land, sea, cyber, and space).
JTF (Joint Task Force)	A temporary or standing military command composed of elements from multiple services organized to perform a specific mission.
NDAA (National Defense Authorization Act)	Annual federal legislation specifying the DoD's budget, policies, and programs.

Term / Acronym	Definition
NORAD (North American Aerospace Defense Command)	A binational U.S.-Canadian organization providing aerospace warning and control for North America.
NORTHCOM (U.S. Northern Command)	U.S. Combatant Command responsible for homeland defense, civil support, and security cooperation within North America.
NOTAM (Notice to Air Missions)	An FAA communication advises pilots of temporary airspace restrictions or hazards.
PART 107 (Small UAS Rule)	The FAA regulation establishes standards for commercial drone operations in U.S. airspace, including pilot certification, visual line-of-sight requirements, and operational limits.
REMOTE ID	FAA requirement that drones broadcast identification, location, and operator information while airborne, serving as a "digital license plate."

Term / Acronym	Definition
RULE OF ENGAGEMENT (ROE)	Directives that define circumstances and limitations under which forces may initiate or continue combat engagement.
SOP (Standard Operating Procedure)	Written guidance outlining routine or emergency procedures to ensure consistency and compliance.
SRD (Shared Responsibility Doctrine)	The framework introduced in *"The Sky Isn't Empty"* emphasizes the collective responsibility of citizens, institutions, and governments for national resilience.
STEM (Science, Technology, Engineering, and Mathematics)	Academic disciplines are essential to innovation and are often integrated with civic education in modern curricula.
TFR (Temporary Flight Restriction)	A short-term FAA airspace restriction issued to protect people or property during special events, disasters, or national security operations.

Term / Acronym	Definition
TRUST (The Recreational UAS Safety Test)	An FAA-mandated online test verifying basic knowledge for recreational drone pilots.
UAS (Unmanned Aircraft System)	The collective term for a drone and its components, including the aircraft, control station, and communications links.
UNCLASSIFIED / CLASSIFIED LEVELS	Information designations based on security sensitivity; determine clearance and access protocols.
USC (United States Code)	Compilation of permanent federal laws organized by subject matter.
VISUAL LINE OF SIGHT (VLOS)	The operational requirement is that a drone pilot must maintain continuous unaided visual contact with the aircraft.
VTOL (Vertical Takeoff and Landing)	Aircraft capability to ascend, hover, and descend vertically without runways, common in many modern drone systems.

End State

Every policy, SOP, and training program developed from this work should utilize a shared language. **Clarity creates unity.** When citizens, engineers, and commanders use the same terms, they build not but also trust.

"Vocabulary is the architecture of understanding. Without it, coordination collapses.", CSM Sheldon A. Watson.

Appendix G: References and Source Notes

"Facts anchor conviction; truth guides command.", CSM Sheldon A. Watson.

Section I: Academic and Theoretical Sources

Aguirre, B. E. (2022). *Adaptive Systems and Social Resilience: A Framework for Public Administration. Public Administration Review, 82*(1), 10–24. https://doi.org/10.1111/puar.13325

Anderson, J., & Rainie, L. (2023). *The Future of AI and Ethics: Human Oversight in an Automated World.* Pew Research Center.

Boin, A., & Lodge, M. (2016). *Designing Resilient Institutions: Leadership and Crisis Governance in Public Administration. Governance, 29*(4), 465–480.

Creswell, J. W., & Poth, C. N. (2018). *Qualitative Inquiry & Research Design: Choosing Among Five Approaches* (4th ed.). SAGE Publications.

Folke, C., Carpenter, S. R., Walker, B., Scheffer, M., Chapin, F. S., & Rockström, J. (2021). *Resilience Thinking: Integrating Social-Ecological Systems in Governance. Annual Review of Environment and Resources, 46*(1), 1–26.

Lipsky, M. (2010). *Street-Level Bureaucracy: Dilemmas of the Individual in Public Services.* Russell Sage Foundation.

Moore, M. H. (2022). *Creating Public Value in a Networked Age.* Harvard Kennedy School Working Paper Series.

Yin, R. K. (2022). *Case Study Research and Applications: Design and Methods* (7th ed.). SAGE Publications.

Section II: Government, Policy, and Technical Documents

Congressional Research Service. (2025). *Department of Defense Counter-Unmanned Aircraft Systems (C-UAS): Authorities and Oversight.* CRS Report R48477.

Cybersecurity and Infrastructure Security Agency. (2024). *CISA Counter-UAS Strategy: Safeguarding the National Airspace.* U.S. Department of Homeland Security.

Department of Defense. (2023). *Directive 3000.09: Autonomy in Weapon Systems.* Washington, DC: Office of the Secretary of Defense.

Federal Aviation Administration. (2025). *Summary of Small Unmanned Aircraft Regulations (Part 107) and Remote Identification Requirements.* Retrieved from https://www.faa.gov/uas

Federal Aviation Administration. (2024). *Advisory Circular 91-57B: Exception for Limited Recreational Operations of Unmanned Aircraft.* U.S. Department of Transportation.

Federal Communications Commission. (2023). *Spectrum Policy for Emerging Technologies: Managing Electromagnetic Congestion.* FCC Technical Bulletin.

Government Accountability Office. (2024). *Homeland Defense: DOD Needs a Comprehensive Strategy to Counter Small UAS Threats (GAO-24-106158)*. Washington, DC.

North American Aerospace Defense Command (NORAD) & U.S. Northern Command (USNORTHCOM). (2023). *Homeland Defense Integration and Air Domain Awareness Posture Review*. Peterson Space Force Base, CO.

U.S. Department of Homeland Security. (2023). *Strategic Plan FY 2023–2027*. Washington, DC.

U.S. Department of Justice. (2023). *Counter-Unmanned Aircraft Systems Legal Authorities and Limitations Brief*. Washington, DC.

U.S. Office of Science and Technology Policy. (2024). *Blueprint for an AI Bill of Rights: Making Automated Systems Work for the American People*. Executive Office of the President.

Section III: Media, Case Studies, and Real-World Incidents

Baker, M. (2023, November 12). *Drones Disrupt Firefighting Operations in California Wildfires, the Los Angeles Times*.

Brumfield, E. (2022, September 3). *Drones Over Stadiums: The New Security Challenge for Large-Scale Events. Homeland Security Today*.

Garcia, R. (2024, March 15). *Phoenix Elementary Drone Incident Raises Questions About Automation and Accountability*, The Arizona Republic.

McLeod, S. (2023, April 2). *Drone Hobbyist Pleads Guilty After Interfering with Aerial Fire Response*. The Denver Post.

Reuters Staff. (2023, December 19). *Cartel-Operated Drones Increase Cross-Border Surveillance Capabilities*. Reuters Defense & Security.

Wheeler, J. (2024, July 11). *AI in the Loop: How Local Governments Are Managing Automation in Decision-Making*. Governing Magazine.

Section IV: Faith, Leadership, and Ethical Foundations

Aquinas, T. (1265–1274/1947). *Summa Theologica* (Fathers of the English Dominican Province, Trans.). Benziger Bros.

Lewis, C. S. (1943). *The Abolition of Man*. HarperCollins.

Niebuhr, R. (1944). *The Children of Light and the Children of Darkness: A Vindication of Democracy and a Critique of Its Traditional Defense*. Charles Scribner's Sons.

Roosevelt, F. D. (1937). *Inaugural Address*. National Archives.

Scripture , Proverbs 4:7 (NIV): *"Wisdom is the principal thing; therefore get wisdom: and with all thy getting get understanding."*

Watson, S. A. (2025). *Faith as Framework: Applying Scriptural Principles to Modern Defense*. Unpublished manuscript.

Notes on Citations and Integration

All quotations within the manuscript that include *"CSM Sheldon A. Watson"* are original to the author and should be cited as personal or creative attributions within the publication.

Federal statutes referenced throughout (e.g., 10 U.S.C. § 130i; 49 U.S.C. § 44809) align with the current versions of the United States Code as of FY 2025.

Where available, hyperlinks and official document titles have been maintained for professional verification and accessibility.

End State: Intellectual Integrity in Motion

A credible argument does not rest on authority; it rests on accountability to truth.

These sources collectively ensure that *The Sky Isn't Empty* remains not only persuasive but also anchored, verified, and enduring.

"Knowledge without wisdom builds machines; wisdom without knowledge builds myths. Leadership must balance both." CSM Sheldon A. Watson.

"Awareness without action is observation. Action without ethics is chaos.", CSM Sheldon A. Watson.

The Sky Isn't Empty: A Call to the American Sentinel

I am a citizen of a connected nation, a guardian of my community, and a witness to the sky above me.

I will not surrender awareness for comfort, nor trade vigilance for convenience.
I understand that freedom without stewardship. Is fragility disguised as liberty.

I will question systems that think for me, challenge technology that erodes accountability, and defend the space between innovation and integrity.

I will engage, not complain. Prepare, not panic. Inform, not inflame. Lead, not follow.

I will teach the next generation that security begins not in secrecy, but in shared responsibility.

When drones fill the sky, I will fill the silence with truth. When autonomy advances, I will advance with conscience. When leadership falters, I will stand, calm, deliberate, and faithful.

Because the sky is not empty, it reflects the soul of the nation that built it. Signed, S.A.W

About the Author

CSM Sheldon A. Watson (USA) serves as Command Sergeant Major for the United States Army. With over two and a half decades of service, his career spans combat leadership, homeland defense, and public administration. A doctoral candidate at Liberty University, he is devoted to advancing responsible innovation, ethical governance, and civic resilience.

He lives in Colorado Springs with his wife, Soleil, and their sons, Silas and Shane.